UNITEXT - La Matematica per il 3+2

Volume 108

More information about this series at http://www.springer.com/series/5418

Pablo Pedregal

Optimization
and Approximation

 Springer

Pablo Pedregal
Department of Mathematics
Universidad de Castilla-La Mancha
Ciudad Real
Spain

ISSN 2038-5714   ISSN 2532-3318 (electronic)
UNITEXT - La Matematica per il 3+2
ISSN 2038-5722   ISSN 2038-5757 (electronic)
ISBN 978-3-319-64842-2   ISBN 978-3-319-64843-9 (eBook)
DOI 10.1007/978-3-319-64843-9

Library of Congress Control Number: 2017949160

Printed on acid-free paper

This Springer imprint is published by Springer Nature
The registered company is Springer International Publishing AG
The registered company address is: Gewerbestrasse 11, 6330 Cham, Switzerland

Preface

There seems to be an ever-increasing need to get to the basic mathematics behind applications as quickly as possible. Optimization is one such body of ideas that permeates all of Science and Engineering. Quite often, interested readers who look at optimization ideas in a utilitarian way find that they need to dive into textbooks requiring a lot of formal training in order to find the basic concepts and techniques that they are seeking. This is quite frustrating at times, for all they require is a well-founded and educated intuition that will endow them with a basic under-standing and some affordable computational tools to tackle problems in practice. In addition, even if a person is solely interested in a particular body of ideas in Optimization, he/she may need to know a little of everything so as to gain an overall mental picture. Real-life applications may demand identification of the nature of problems, the framework in which they can be treated, some simple computational methods to deal with equivalent toy models, etc. Once this initial fundamental information has been appropriately defined, further specialized literature may have to be sought. This text aims, as one way among various possibilities, to endow readers with such a basic overall background in Optimization that may enable them to identify a problem as a mathematical program, linear or nonlinear, or as a continuous optimization situation, in the form of either a variational problem or an optimal control case. It is also my experience that students who discover that it is virtually impossible to solve by hand even the more innocent-looking optimization problems become frustrated when they realize that in order to play with approxi-mated solutions of simple variations of those situations, they will need to wait until much more has been learned about how to approximate optimal solutions. To propose one basic, systematic, and coherent solution for this approximation issue across all main areas of Optimization is also a main motivation for this text. It is just one affordable possibility among many others, with no intention of competing with more sophisticated and accurate methods in the computational arena. I describe some simple, yet non-entirely trivial, procedures to approximate optimal solutions of easy problems so that students may experience the joy of seeing with their own eyes optimal solutions of problems. There are, as just pointed out, other good and reasonable possibilities and solutions for this computational issue.

I feel this book is unique in that it makes an attempt to integrate all those main fields that share sufficient features to be placed under the umbrella of Optimization, a huge area of work in which many people are making progress on a daily basis. But there are so many distinct elements among those subareas that it is scarcely possible to make a deep analysis, within a single text, that covers all of them. In this sense, our treatment is basic and elementary but, as already pointed out in the preceding paragraph, it seeks to draw an overall picture of the interconnection among the main parts of Optimization that may be helpful to students.

The book will be of help to students of Science and Engineering, at both undergraduate and graduate level, who for the first time need to acquaint themselves with the basic principles of all areas of Optimization. In particular, the book ought to suffice for the content of typical courses in Science and Engineering majors. Familiarity with basic linear algebra, multivariate calculus, and basic differential equations is a fundamental prerequisite. Senior researchers whose main field of expertise is not Optimization, but who need to understand the basic theory and eventually to use approximation techniques for the problems in which they are interested, may also find these pages useful. I warn experts in some of the areas briefly covered in the text that they may find the coverage a little disappointing since the treatment of each area is too basic to be of interest for specialists. Some material or concepts that are regarded as fundamental in some subfields of Optimization may be missing from this book. Let me again stress that my intention has been to provide one basic, cross-cutting source relevant to all important areas in Optimization. This obviously means that it is not possible to attain great depth within individual chapters.

The text may serve several purposes:

- It can enable the user to avoid all issues about approximation techniques and to focus on basic analytical ideas in all chapters.
- It can enable the user to concentrate on mathematical programming problems, including issues about approximation, either by using the ideas in the text concerning simulation in practice or by utilizing other software packages.
- It may be used as the basis for a course focusing on continuous optimization problems, including approximation techniques.
- It can serve other specific purposes when time is scarce and there is a need to concentrate on narrower objectives.

It may be taken as the basic textbook for an Optimization course for any Science or Engineering major, and also for Master's courses. The text has a modular structure in the sense that even separate individual sections may serve as short introductions to specific topics or techniques, depending on needs.

I have paid a lot of attention to exercises as I believe them to be a fundamental part of a text aiming to support basic learning in any field of mathematics. New ideas and motivations have been associated with exercises in an attempt to make them more inspiring and enlightening. I have divided the exercises into three distinct categories: those supporting main analytical concepts, those aiming to provide basic training, and those of a more advanced nature. About 160 exercises with full solutions are gathered in the final chapter. Most of these exercises have been tried out in the Optimization course that the author and other colleagues have taught in the Math Dep of UCLM over the past 15 years. In this regard, particular thanks go to E. Aranda, J.C. Bellido and A. Donoso. I do not claim to have gathered all possible problems for such an introductory course. I could hardly do so, as there are hundreds of examples scattered throughout the literature. My goal, rather, has been to propose enough exercises to ensure that they enable the reader to achieve a reasonably mature grasp of the main concepts I have tried to convey. Whenever possible, I have made an attempt to inspire further investigation through some of them. I have given the source of problems whenever known, but I cannot guarantee to have done so in a systematic way; thus, readers may know other sources for some of the problems that have not been indicated in the text. Many of the problems can be found in other textbooks, so I do not claim the slightest originality here. Needless to say, anyone using this book as a source text can propose many other exercises to students. As a matter of fact, many other exercises can be found in the bibliography I have selected.

Concerning references, I have not tried to be exhaustive, and some interesting resources are probably missing. My aim has been to provide a limited number of well-chosen items so as to encourage further investigation or to identify where to look for deeper answers to more advanced questions. Each part has its own final bibliography section, though some are common to several parts.

I would also like to take this opportunity to thank the members of the Editorial Board of the series UNITEXT, and particularly A. Quarteroni, for their help and assistance in the reviewing process for the manuscript and for suggesting very appropriate improvements which have led to this final form. My gratitude also goes to Francesca Bonadei and Angela Vanegas from Springer, whose constant support made the whole process an easy journey.

Ciudad Real   Pablo Pedregal
June 2017

Contents

About the Author

Pablo Pedregal studied Mathematics at U. Complutense (1986) and received his Ph.D. degree from the University of Minnesota at the end of 1989, under the guidance of D. Kinderlehrer. In 1990, he became Associate Professor at U. Complutense. During the academic year 1994–95 he moved to the young U. de Castilla-La Mancha, and in 1997 he won a full professorship. His field of interest focuses on variational techniques applied to optimization in a broad sense, including, but not limited to, calculus of variations—especially vector, non-convex problems, optimal design in continuous media, optimal control, etc, and more recently he has explored variational ideas in areas like controllability, inverse problems, PDEs, and dynamical systems. Since his doctorate, he has regularly traveled to research centers in the USA and Europe. He has written more than one hundred research articles, and several specialized books.

Chapter 1
Overview

1.1 Importance of Optimization Problems

The term optimization suggests good feelings from its usual meaning in everyday life. At a more precise, scientific level, even to the beginner, it sounds as an important field in the decision-making process in many areas of science and engineering. But when one starts to see what kind of situations can be treated and analyzed, even at a purely academic level, and extrapolate its applicability to more realistic problems, then there is no question that we are before a fundamental area, and the better we understand its principles and techniques, the more valuable our decisions will be.

This short text is but a timid introduction to this important subject, written in a way so that students are led directly to the point, and the main issues of mathematical programming problems, on the one hand, and of variational and optimal control problems, on the other, are treated and explained. It is therefore like a first step. As one becomes interested in deepening his or her knowledge and maturing the techniques, other more profound and elaborate sources will have to be studied.

One main important aspect of optimization problems that seem to require a lot of attention in a first exposure is to understand the process of going from the statement (most probably incomplete) in plain words of what one would like to design or the decision to be made, to the precise formulation in terms of variables, costs, constraints, parameters, etc. If this process is not resolved fully and accurately, what comes after that may be irrelevant altogether, and more often than not, wrong, or pointless. To feel some confidence in going through this passage is a personal endeavor facilitated through the working-out process of many practical problems and situations. There is no substitute for this. The material in this book, however, aims at a different objective: understanding the importance of necessary conditions of optimality, and structural assumptions to ensure sufficiency of those conditions to find the desired optimal solutions. We hope to bring readers to a willingness to deepen their knowledge further by looking for more advanced sources.

Another fundamental aspect of optimization relates to finding optimal solutions in practice. Our readers will soon realize that this is a dramatic need as very few

© Springer International Publishing AG 2017

P. Pedregal, *Optimization and Approximation*, UNITEXT 108,

DOI 10.1007/978-3-319-64843-9_1

and only well-prepared examples can be treated by naked hands. An affordable and efficient tool to approximate optimal solutions, even for academic problems, is almost mandatory. We have proposed one consistent and unifying such possibility. This is, we believe, one of the main strengths that we have tried to implement in this booklet. The title aims at stressing this issue. There are, however, many other alternatives.

1.2   Optimization

In every optimization problem or situation, there is always two preliminary steps to be covered, and another two main steps towards the solution.

The two preliminary steps are:

1. translate, into precise equations and formulas, the various ingredients of the example like the choice of suitable variables, the distinction between variables and parameters, the form of the objective functional, the restrictions to be respected, whether they come in the form of equalities or inequalities, etc.;
2. an important preliminary analysis should lead us to conclude if the solution we propose to seek does indeed exist, or whether the cost functional could decrease indefinitely maintaining feasibility, in case we should look for a minimum; or whether the cost functional could take larger and larger values without violating admissibility.

If we have succeeded in completing the two previous steps, and convince ourselves (and others) that what we seek is indeed there, waiting for us to be found, there are another two main steps to be examined:

1. detect the possible candidates for the extreme value we are searching for through the so-called optimality conditions;
2. check if especial structural conditions on the various functions involved ensure that optimality conditions are sufficient for finding what we seek.

This text aims at being a first contact with these two aspects of optimization problems: examine, understand, and manipulate optimality conditions; and start to appreciate the importance of those special properties that guarantee that necessary conditions of optimality are, indeed, sufficient to find optimal solutions.

1.2.1   Unconstrained Optimization and Critical Points

The paradigm of an optimization problem is to

$$\text{Minimize in } \mathbf{x} \in \mathbb{R}^N : \quad f(\mathbf{x}), \tag{1.1}$$

where the function $f : \mathbb{R}^N \to \mathbb{R}$ is given. We seek a particular point $\bar{\mathbf{x}}$ where the cost function f attains its global minimum, so that we are sure that there is no other $\tilde{\mathbf{x}}$ that beats the one we have found. Such an $\bar{\mathbf{x}}$ is called an optimal solution of the problem. The whole point of mathematical programming is to find (compute approximately) those points $\bar{\mathbf{x}}$. This problem (1.1) is called unconstrained because vectors $\mathbf{x}$ may run freely, in an unrestricted fashion, throughout $\mathbb{R}^N$.

From Calculus, we know that the global minima of a certain function like f above need to be critical points of f. But we are also well aware that there may be many more critical points than just those we are interested in. At any rate, the (non-linear) system $\nabla f(\mathbf{x}) = \mathbf{0}$ is like a very good place to start looking for those ambitioned solutions. There are, hence, like two main steps in finding those optimal solutions. First, all critical points of f become candidates where the (global) minimum and/or maximum may hide. Secondly, a sensible discernment is mandatory to sort out the true extreme of the function; or else, check if additional properties of f ensure that critical points correspond to the global minimum sought. The critical-point system

$$\nabla f(\mathbf{x}) = \mathbf{0}$$

corresponds to optimality conditions for this paradigmatic case, while the convexity of f accounts for that special property ensuring that all critical points will be global minimizers.

1.2.2 Constrained Optimization

One of the headaches, and of its importance at the same time, of mathematical programming is precisely the situation where the set of competing vectors $\mathbf{x}$ is restricted or constrained through various equations in the form of equality or inequality, and how the presence of these affects the process of finding optimal solutions. Roughly speaking, our model problem for mathematical programming will be

$$\text{Minimize in } \mathbf{x} \in \mathbb{R}^N : \quad f(\mathbf{x})$$

but only among those vectors $\mathbf{x} \in \mathbb{R}^N$ which comply with the conditions

$$\mathbf{g}(\mathbf{x}) \leq \mathbf{0}, \quad \mathbf{h}(\mathbf{x}) = \mathbf{0},$$

for vector-valued functions $\mathbf{g}$ and $\mathbf{h}$. How does the presence of these constraints expressed through $\mathbf{g}$ and $\mathbf{h}$ affect necessary conditions of optimality? What role does convexity play for sufficiency of those same conditions? For these constrained problems, optimality conditions are universally referred to as the Karush–Kuhn–Tucker (KKT) optimality conditions, and its general form is

$$\nabla f(\mathbf{x}) + \mathbf{y}\nabla \mathbf{h}(\mathbf{x}) + \mathbf{z}\nabla \mathbf{g}(\mathbf{x}) = \mathbf{0},$$

$$\mathbf{z} \cdot \mathbf{g}(\mathbf{x}) = 0, \quad \mathbf{h}(\mathbf{x}) = \mathbf{0},$$

$$\mathbf{z} \geq \mathbf{0}, \quad \mathbf{g}(\mathbf{x}) \leq \mathbf{0},$$

for (unknown) vectors of multipliers $\mathbf{y}$ and $\mathbf{z}$, in addition to $\mathbf{x}$, of the appropriate dimensions. When all functions involved in the problem, f and all components of $\mathbf{g}$ and of $\mathbf{h}$, are linear, the problem is identified as a linear-programming problem. They enjoy very special properties.

1.2.3 Variational Problems

Variational problems, and optimal control problems as well, are optimization problems of a very different nature. No wonder that they are often treated as a separate area. The most striking difference is that feasible objects modeling strategies are described through functions instead of vectors. In this way, equations of optimality come in the form of differential equations, and costs are typically expressed as integrals instead of functions. To stress this main difference, functions in this context are referred to as functionals. The usual necessary conditions of optimality are universally known as the Euler–Lagrange equations.

Our model problem will be

$$\text{Minimize in } u(t): \quad \int_0^T F(t, u(t), u'(t))\, dt$$

under $u(0) = u_0, u(T) = u_T$. Here the integrand $F(t, u, p) : [0, T] \times \mathbb{R} \times \mathbb{R} \to \mathbb{R}$, the upper limit T, and the two numbers u_0, u_T make up the full data set of the problem. Optimal solutions $\overline{u}$, when they exist, are to be found as a solution of the underlying Euler–Lagrange equation which adopts the form

$$-[F_p(t, \overline{u}(t), \overline{u}'(t))]' + F_u(t, \overline{u}(t), \overline{u}'(t)) = 0 \text{ in } (0, T),$$

together with the end-point conditions $u(0) = u_0, u(T) = u_T$. This will be the main subject in the second part of the book.

1.2.4 Optimal Control Problems

There are some very special ingredients in every optimal control problem. They share, at the same time, many features with variational problems so that it is not rare to find both areas treated in the same textbook. There are here typically two sets of variables. The so-called state variables specifying where the system we are interested in stands for each time, and the control variables through which we can interact, manipulate, and control the system to our advantage. The state law is probably the most important characteristic element in every optimal control problem.

The fundamental paradigmatic situation we will be concerned with is to

$$\text{Minimize in } u(t): \quad \int_0^T F(x(t), u(t))\, dt$$

under the conditions

$$x'(t) = f(x(t), u(t)) \text{ in } (0, T), \quad x(0) = x_0, u(t) \in K,$$

where T, the integrand F, the right-hand side of the state law f, the initial condition x_0, and the subset $K \subset \mathbb{R}$ make up the data set of the problem. Here, Pontryaguin's principle is the main form of optimality conditions. If we put

$$H(u, x, p) = F(x, u) + p\, f(x, u),$$

called the Hamiltonian of the problem, then optimal solutions $\overline{u}$ go with optimal state $\overline{x}$, and optimal costate $\overline{p}$, and they all together need to comply with the following conditions

$$\overline{p}'(t) = -\frac{\partial H}{\partial x}(\overline{u}(t), \overline{x}(t), \overline{p}(t)), \quad \overline{x}'(t) = \frac{\partial H}{\partial p}(\overline{u}(t), \overline{x}(t), \overline{p}(t)),$$

in the full interval $(0, T)$, as well as

$$H(\overline{u}(t), \overline{x}(t), \overline{p}(t)) = \min_{u \in K} H(u, \overline{x}(t), \overline{p}(t)).$$

Initial condition $\overline{x}(0) = x_0$, and transversality condition $\overline{p}(T) = 0$ are to be respected too. This will be our main theme in the third part.

1.3 Approximation

Even for simple examples with a rather moderate or low number of variables, finding optimal solutions may be an impossible task. Without some affordable methods to approximate solutions of non-prepared examples, we are left with a sense of helplessness as the problems that can be solved by using directly optimality conditions by naked hands is very limited. Approximation techniques are then to be examined from the very beginning. This is one main emphasis in this booklet, as the title tries to convey. We propose methods to approximate optimal solutions for all types of situations, and illustrate the performance of them by dealing with selected examples. Fortunately, these numerical procedures are not too sophisticated, though they are not completely trivial. We do not intend to get into a more delicate discussion about specialized numerical methods for each category of problems. These are reserved for the experts when very accurate approximations are necessary.

The basis for all of our approximation schemes is just to be able to find good approximations for unconstrained problems. Indeed, if one can count on some sort of subroutine to approximate efficiently the solutions of unconstrained cases, then one can very easily use it to find optimal solutions for constrained problems, by iterating a collection of unconstrained problems where various elements of it are updated iteratively until convergence.

In simple terms, we will have a master function $L(\mathbf{x}, \mathbf{y})$ depending on two sets of variables. The basic iterative step will consists in two tasks: first, minimize in the variable $\mathbf{x}$, for given $\mathbf{y}$, but in an unrestricted way; secondly, update $\mathbf{y}$ by using the previously computed minimizer. To be more specific, consider the problem of

$$\text{Minimize in } \mathbf{x} \in \mathbb{R}^N : \quad f(\mathbf{x})$$

under those feasible vectors $\mathbf{x}$ that comply with

$$g(\mathbf{x}) \leq 0,$$

where both f and g are smooth functions. In this particular situation, variable $\mathbf{y}$ would be just one dimensional because we would have just one inequality constraint to respect associated with the function g. We would put

$$L(\mathbf{x}, y) = f(\mathbf{x}) + e^{yg(\mathbf{x})}.$$

The main iterative step to approximate the optimal solution of the restricted problem, would consists of two main parts:

1. for given $y \geq 0$, solve the unrestricted problem

$$\text{Minimize in } \mathbf{x} : \quad L(\mathbf{x}, y),$$

and let $\mathbf{x}_y$ be the (approximated) solution;
2. if this solution is not yet a good approximation, update y by the rule

$$y \mapsto y e^{yg(\mathbf{x}_y)},$$

and go back to the first step.

Whenever this iterative procedure converges, it will do so to the (one) (local) optimal solution of the constrained problem.

For the case of variational and optimal control problems the perspective is exactly the same, but its implementation in practice requires more work. On the one hand, we need to learn how to approximate optimal solutions of variational problems with no additional restrictions, except perhaps end-point conditions. The practical difficulty, for constrained problems, is that there will be additional functions defining the unrestricted variational problem whose solutions have to be calculated at each iterative step. Once this has been overcome, the procedure can be efficiently used to approximate the optimal solutions under all kinds of constraints.

1.4 Structure and Use of the Book

We have divided the book into three parts: mathematical programming, variational problems, and optimal control. As has been stressed above, there are good reasons for this division. There is a final chapter in each part focused on numerical approximation, and one or two where a first treatment of each type of optimization problem is exposed.

The text is addressed to undergraduate students who would like to understand the basics of mathematical programming, variational problems, and/or optimal control. It can also serve to students in science and engineering willing to understand what optimization is about, and have a first glimpse of this fascinating universe. It can serve to several purposes, as indicated in the Preface:

- as a source for a first course in mathematical programming with a special emphasis both in analytical ideas and simulation;
- for a course on optimization with no particular concern about approximation, but covering mathematical programming, variational problems, and optimal control;
- as the basic reference for an introductory course to variational problems and optimal control with some interest in the numerical simulation.

Prerequisites reduce to a basic knowledge of Multivariate Calculus, Linear Algebra, and Differential Equations.

Our explicit intention has been to avoid mathematical rigor in terms of formal proofs, most possible general set of assumptions, and neat analysis, when this would require an unreasonable emphasis on analytical tools or techniques given the scope of the text. Though we have tried to state results and facts as accurately as the general tone of the text permits, many important issues have been deliberately left out. On the other hand, special care has been taken concerning exercises, and their solutions, as this is an unavoidable part of the learning process. References are provided to pursue further the analysis of mathematical tools for optimization problems, approximation techniques, realistic cases, etc.

Part I
Mathematical Programming

Chapter 2
Linear Programming

There are various important features exclusive of linear programming without counterpart in non-linear programming. In this chapter we focus on those, and aim at a first, basic understanding of this special class of optimization problems.

2.1 Some Motivating Examples

As a way to clearly see and understand the scope of linear programming, let us examine three typical situations. We will mimic the passage from the statement of the problem in common language to its precise formulation into mathematical terms. This transformation is as important as the later solution procedure or approximation, to the extent that failure at this step will translate into a waste of time and/or resources. Statements come in a rather typical form of linear programming problems.

Example 2.1 A certain company is willing to invest 75,000 million € in the purchase of new machinery for its business. It can choose among three models A, B, and C, which differ in several features, including price. It is estimated that model A will produce a annual benefit of 315 million €, while the price per unity is 1500 million €; for model B, there will be an estimated benefit of 450 million €, while the price per unity is 1750 million €; likewise for C with 275, and 1375 million €, respectively. Those machines also have requirements and expenses concerning maintenance in terms of facilities and labor; specifically, type A needs 20 million € for labor each, and 34 days per year for maintenance facilities; type B, 27 million € for labor, and 29 days; and type C, 25 million € for labor, and 24 days. There is a maximum allowance of 1000 million € for maintenance purposes, and 800 days for availability of facilities per year. Make a recommendation about how many machines of each type the company should purchase so as to maximize yearly benefits.

This is one of those examples where the information conveyed by the statement can be efficiently recorded in a table where one can clearly see the main ingredients of a linear programming problem.

© Springer International Publishing AG 2017
P. Pedregal, *Optimization and Approximation*, UNITEXT 108,
DOI 10.1007/978-3-319-64843-9_2

Type	A	B	C	Resources
Price	1500	1750	1375	75000
Labor	20	27	25	1000
Days	34	29	24	800
Benefit	315	450	275	

Upon reflecting on the content of this table, one realizes that we need to

$$\text{Maximize in } (x_A, x_B, x_C): \quad 315x_A + 450x_B + 275x_C$$

under the constraints

$$1500x_A + 1750x_B + 1375x_C \leq 75000,$$
$$20x_A + 27x_B + 25x_C \leq 1000,$$
$$34x_A + 29x_B + 24x_C \leq 800.$$

Here variables x indicate number of machines of each type to be purchased. We should also demand, for obvious reasons, that $x_A, x_B, x_C \geq 0$, and even that they ought to take on integer values.

Example 2.2 A scaffolding system. Figure 2.1 represents a system where the sum of the maximum weights of the two black boxes is to be maximized avoiding collapse. Lengths are given in the picture, and cables A, B, C, and D can withstand a maximum tension 200, 100, 200, 100, respectively, in the appropriate units. We should express equilibrium conditions for forces and momenta for each of the two rods. Namely, if

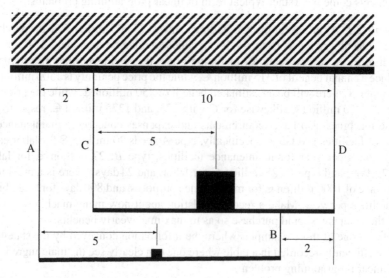

Fig. 2.1 A scaffolding system

we set x_1 and x_2 for the loads of the lower and upper rods, respectively, and T_A, T_B, T_C, T_D the tensions in their respective cables, we must have

$$x_1 = T_A + T_B, \quad x_2 + T_B = T_C + T_D,$$
$$5x_1 = 10T_B, \quad 5x_2 + 8T_B = 10T_D.$$

Therefore the problem may be written in the form

$$\text{Maximize in } (x_1, x_2): \quad x_1 + x_2$$

subject to

$$x_1 = T_A + T_B, \quad x_2 + T_B = T_C + T_D,$$
$$5x_1 = 10T_B, \quad 5x_2 + 8T_B = 10T_D,$$
$$T_A \leq 200, \quad T_B \leq 100, \quad T_C \leq 200, \quad T_D \leq 100.$$

This format can be simplified if we use equilibrium conditions to eliminate the tensions T_X, $X = A, B, C, D$. Indeed, it is elementary to find

$$T_A = T_B = \frac{x_1}{2}, \quad T_C = \frac{x_1}{10} + \frac{x_2}{2}, \quad T_D = \frac{2x_1}{5} + \frac{x_2}{2},$$

and imposing the maximum allowable values of these loads, we are led to the constraints

$$x_1 \leq 200, \quad x_1 \leq 400, \quad x_1 + 5x_2 \leq 2000, \quad 4x_1 + 5x_2 \leq 1000. \tag{2.1}$$

The problem now becomes to maximize $x_1 + x_2$ under (2.1). It is, however, interesting to realize that restrictions in (2.1) can be simplified even further because some constraints are more restrictive than others. In the $x_1 - x_2$ plane, it is easy to find that constraints can be reduced as it is indicated in the final form of the problem

$$\text{Maximize in} (x_1, x_2): \quad x_1 + x_2$$

under

$$0 \leq x_1 \leq 200, \quad 0 \leq x_2, \quad 4x_1 + 5x_2 \leq 1000.$$

Example 2.3 A small electric power supply company owns two generators. The first one yields a net benefit of $3\,€$/MWh with a maximum exit power of 4 MWh, while the second yields a benefit of $5\,€$/MWh with a maximal exit power of 6 MWh. Cooling constraints demand that three times the exit power of the first plus twice that of the second cannot exceed, under no circumstances, 18 MWh. The issue is to maximize the global net income coming from the two generators, and the corresponding optimal working conditions. This example is easy to formulate. If we let x_1 and x_2 be the exit powers of the two generators, respectively, to be determined, the statement clearly translate into the problem

$$\text{Maximize in } (x_1, x_2): \quad 3x_1 + 5x_2$$

subject to

$$0 \le x_1 \le 4, \quad 0 \le x_2 \le 6, \quad 3x_1 + 2x_2 \le 18.$$

Sometimes being able to answer additional questions is of paramount practical importance. In the context of this particular problem, suppose that the company is considering the possibility of expanding the maximal exit power of one of the two generators in order to produce a rise in net income. Which one is the best for a maximum such increment in benefits? What would the expected benefit be if the company would increase in 1 MWh the exit power of such a generator?

There are very classical LP examples that almost every text in Mathematical Programming treats: the diet problem, the transportation problem, the portfolio situation, etc. We will also describe some of them later.

2.2   Structure of Linear Programming Problems

As soon as we ponder a bit on the examples described in the preceding section, we immediately realize that a linear programming problem is one in which we want to

$$\text{Optimize in } \mathbf{x} \in \mathbb{R}^N: \quad \mathbf{c} \cdot \mathbf{x} \quad \text{under} \quad \mathbf{Ax} \le \mathbf{b}, \mathbf{x} \ge \mathbf{0}.$$

The term "optimize" conveys the idea that we could be interested sometimes in maximizing, and some others in minimizing. Constraints could also come in various forms, though we have chosen one in which they often occur. Some of them could also be equalities rather than inequalities. There are then three main ingredients to determine one such linear programming problem:

1. vector $\mathbf{c} \in \mathbb{R}^N$ is the cost vector;
2. vector $\mathbf{b} \in \mathbb{R}^n$ is the constraint vector;
3. matrix $\mathbf{A} \in \mathbb{R}^{n \times N}$ is the constraint matrix.

The important issue is to understand that:

The distinctive feature of every linear programming problem is that every function involved is linear.

Even the sole occurrence of one non-linear function (like an innocent square) among ten million linear functions suffices to discard the problem as a linear programming problem.

With those three ingredients, two main elements are determined. One is the cost function itself, which is the inner product (a linear function) $\mathbf{c} \cdot \mathbf{x}$. There is no much to be said about it. The other main element is the so-called feasible set, which incorporates all of the vectors that comply with the constraints

$$\mathbf{F} = \{\mathbf{x} \in \mathbb{R}^N : \mathbf{Ax} - \mathbf{b} \leq \mathbf{0}, \mathbf{x} \geq \mathbf{0}\}.$$

Let us pause for a moment on this feasible set. Three essentially different cases may occur:

1. $\mathbf{F}$ is the empty set. We have placed so many demands on vectors that we have been left with none. Typically, this situation asks for a revision of the problem so as to relax some of the constraint in order to allow some vectors to be feasible.
2. $\mathbf{F}$ is unlimited through, at least, some direction, and so it cannot be enclosed in a ball no matter how big it is.
3. $\mathbf{F}$ is limited or bounded because it can be put within a certain ball.

When the feasible set $\mathbf{F}$ is bounded, it is easy to visualize its structure, especially in dimensions $N = 2$ and $N = 3$. In fact, because constraints are written in the form of inequalities (or equalities) with linear functions, it is not hard to conclude that $\mathbf{F}$ is the intersection of several semi-spaces, or half-spaces. If we consider a single linear inequality $\mathbf{a} \cdot \mathbf{x} \leq b$, with $\mathbf{a} \in \mathbb{R}^N$, $b \in \mathbb{R}$, we see that it corresponds to a half-space in $\mathbb{R}^N$ whose boundary is the hyper-plane with equality $\mathbf{a} \cdot \mathbf{x} = b$. The other half-space corresponds to the other inequality $\mathbf{a} \cdot \mathbf{x} \geq b$. Hence, the feasible set is like a polyhedron in high dimension, with faces, edges, vertices. This is a heuristic statement that conveys in a intuitive way the structure of the feasible set for a linear programming problem. If the feasible set in not bounded, then that polyhedron is not limited through some direction going all the way to infinity.

Let us now examine the problem through the perspective of the solution. For definiteness, consider the linear programming problem

Maximize in $\mathbf{x} \in \mathbb{R}^N$: $\mathbf{c} \cdot \mathbf{x}$ under $\mathbf{Ax} \leq \mathbf{b}, \mathbf{x} \geq \mathbf{0}$. (2.2)

Definition 2.1 A vector $\overline{\mathbf{x}}$ is an optimal solution of (2.2) if it is feasible $\mathbf{A}\overline{\mathbf{x}} \leq \mathbf{b}$, $\overline{\mathbf{x}} \geq \mathbf{0}$, and it provides the maximum possible value of the cost among feasible vectors, i.e. $\mathbf{c} \cdot \overline{\mathbf{x}} \geq \mathbf{c} \cdot \mathbf{x}$ for every feasible vector $\mathbf{x}$.

We may have four different situations:

1. If the problem is infeasible (the admissible set is empty), there is nothing to say, except what we already pointed out earlier. There is no point in talking about solutions here.
2. There is a unique optimal solution. This is, by far, the most desirable situation.
3. There are infinitely many optimal solutions. The point is that as soon as we have two different optimal solution $\bar{\mathbf{x}}_0$, and $\bar{\mathbf{x}}_1$, then we will have infinitely many because every vector in the segment joining those two solutions $t\bar{\mathbf{x}}_1 + (1 - t)\bar{\mathbf{x}}_0$ for arbitrary $t \in [0, 1]$, will again be an optimal solution. This is true because every function involved in a linear programming problem is linear. There is no way for one such problem to have exactly ten, or twenty, or one thousand solutions.
4. There is no optimal solution. This can only happen if the feasible set is unbounded, though some problems with unbounded feasible sets may have optimal solutions. The whole point is that as we move on to infinity through some direction, the cost $\mathbf{c} \cdot \mathbf{x}$ keeps growing or decreasing without limit.

The most important fact for linear programming problems follows.

Proposition 2.1 *Suppose problem* (2.2) *admits an optimal solution. Then there is always one vertex of the feasible set* **F** *which is also optimal.*

Though we have not defined what we understand by a vertex, it will be clear by the following discussion. Consider problem (2.2), and take a feasible vector $\mathbf{x}_0$ (provided the feasible set **F** is not empty). Its cost is the number $\mathbf{c} \cdot \mathbf{x}_0$. We realize that all feasible vectors in the hyperplane of equation $\mathbf{c} \cdot \mathbf{x} = \mathbf{c} \cdot \mathbf{x}_0$ will also yield the same value of the cost. But we are ambitious, and wonder if we could push a bit upward the cost, and still have feasible vectors furnishing that higher value. For instance, we timidly wonder if we could reach the value $\mathbf{c} \cdot \mathbf{x}_0 + \varepsilon$ for a certain small ε. It will be so if the new hyperplane of equation $\mathbf{c} \cdot \mathbf{x} = \mathbf{c} \cdot \mathbf{x}_0 + \varepsilon$ intersects the feasible set somewhere. This new hyperplane is parallel to the first one. We therefore clearly see that we can keep pushing the first hyperplane to look for higher values of the inner product $\mathbf{c} \cdot \mathbf{x}$ until the resulting intersection becomes empty. But there is definitely a last, maximum value in such a way that if we push a little bit further, then we loose contact with the feasible set. A picture in dimension two may help in understanding the situation. See Fig. 2.2 below, and the discussion in Example 2.4. That maximum value is the maximum of the problem, and most definitely in that last moment, there is always a vertex (perhaps more than one vertex if there are infinitely many solutions) which is in the intersection with the feasible set.

Before discussing a specific example where all of the above material is made fully explicit, two further comments are worth stating.

• A same underlying linear programming problem admits many different forms. In some, only inequalities in both directions are involved; in some others, equalities seem to be an important part of the problem; sometimes we seek to maximize a cost; some other times, we look for the minimum, etc. It is easy to go from one given formulation to another, equivalent one. It just requires to know how to change inequalities to equalities, and vice versa. For one thing, an equality is the reunion of two inequalities: $\mathbf{a} \cdot \mathbf{x} = b$ is equivalent to $\mathbf{a} \cdot \mathbf{x} \leq b$, $-\mathbf{a} \cdot \mathbf{x} \leq -b$. On the other hand, passing from an inequality to an equality requires to introduce the so-called slack variables: the inequality $\mathbf{a} \cdot \mathbf{x} \leq b$ is equivalent to $\mathbf{a} \cdot \mathbf{x} + x = b$, $x \geq 0$, where the variable x is the slack associated with the inequality. If we are to respect various inequalities, several slack variables (one for each inequality) need to be introduced. The change from a maximum to a minimum, and vice versa, is accomplished by introducing a minus sign. This is standard.

- When the dimension N of the problem is large, or even moderate, there is no way one can examine by hand the problem to find the optimal solution. If the dimension is low (in particular for $N = 2$), knowing that an optimal solution can always be found in a vertex, it is a matter of examining all of them, and decide on the best one. This idea, based on Proposition 2.1, is the basis for the celebrated SIMPLEX method [22, 29] which for many years has been the favorite algorithm for solving linear programming problems. These days interior point algorithms (see [25], for instance) seem to have taken over. In practice, one needs one such algorithm because real-world problems will always depend on many variables. We will not discuss here this procedure, though there are some true elegant ideas behind, but will be contented with a different numerical scheme to approximate optimal solutions not only for linear programming problems but also for non-linear ones.

Example 2.4 We examine the concrete problem

$$\text{Maximize in } (x_1, x_2) \in \mathbb{R}^2 : \quad x_2 - x_1$$

subject to

$$x_1 + x_2 \leq 1, \quad -x_1 + 2x_2 \leq 2, \quad x_1 \geq -1, \quad -x_1 + 2x_2 \geq -1.$$

Even such a simple example, with a bit of reflection, suffices to realize, at least in a heuristic but well-founded way, the structure of a linear programming problem. We refer to Fig. 2.2 for our discussion.

In the first place, the feasible set is limited by the four lines whose equations are obtained by changing the inequality signs in the constraints to equality. It is very easy to represent those lines in $\mathbb{R}^2$. Those four lines correspond to the four thick continuous lines in Fig. 2.2. The quadrilateral (irregular polyhedron) limited by the four is the feasible set. It is now much easier to realize what we mean by a polyhedron: a set in $\mathbb{R}^N$ limited by hyperplanes.

We now turn to the cost $x_2 - x_1$, and we want to maximize it within our feasible set: find the maximum possible value of the difference of the two coordinates of feasible points. We start asking: are there feasible points with zero cost? Those points would

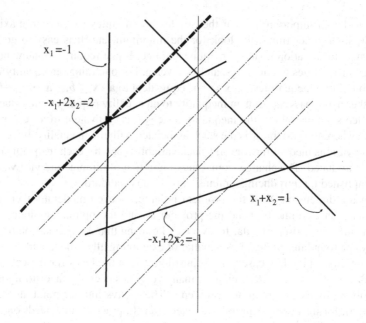

$x_1 = -1$

$-x_1 + 2x_2 = 2$

$x_1 + x_2 = 1$

$-x_1 + 2x_2 = -1$

Fig. 2.2 The feasible set for Example 2.4

correspond to having $x_2 - x_1 = 0$, and this equation represents another straight line in $\mathbb{R}^2$. It is the line with slope one through the origin. We realize that there are feasible points in this line because that line cuts through our quadrilateral. Those points would have zero cost. Since we try to maximize the cost, we start becoming ambitious, and ask ourselves, first rather timidly, if we could go as high as $x_2 - x_1 = 1$. These points would yield cost 1. But we realize that this line is parallel to the previous one but passing through $(1, 0)$. We again have a non-empty intersection with our quadrilateral. Once we understand the game, we become greedy, and pretend to go all the way up to $x_2 - x_1 = 10$. But this time, we have gone too far: there is no intersection of the line $x_2 - x_1 = 10$ with our feasible set. That means that the maximum sought is smaller than 10. But how smaller? At this stage, we notice that the set of parallel lines $x_2 - x_1 = c$ for different values of c represent points with cost c, and so we ask ourselves what the highest value of c is. As we push upward the iso-cost lines $x_2 - x_1 = c$, there is a final such line which touches for the last time our quadrilateral: that is the point, the vertex, where the maximum is achieved, and the corresponding value of c, the maximum.

As one can easily see, simple problems in two independent variables can always be solved graphically as in this last example.

2.3 Sensitivity and Duality

We have seen that, quite often, linear programming problems come in the form

$$\text{Maximize in } \mathbf{x}: \quad \mathbf{c} \cdot \mathbf{x} \quad \text{subject to} \quad \mathbf{A}\mathbf{x} \leq \mathbf{b}, \mathbf{x} \geq \mathbf{0}. \qquad (2.3)$$

Vector $\mathbf{b}$ is typically associated with resources constraints, so that it might be interesting to explore how sensitive the value of the previous maximum is with respect to the specific values of the various components of $\mathbf{b}$. In other words, if $M(\mathbf{b})$ stands for the value of the maximum in the linear programming problem, understood as depending on $\mathbf{b}$ (and also on $\mathbf{A}$ and $\mathbf{c}$ for that matter, though we keep these two ingredients fixed), how does $M(\mathbf{b})$ depend upon $\mathbf{b}$? In particular, and from a practical viewpoint, how do changes on $\mathbf{b}$ get reflected on $M(\mathbf{b})$? Is it worth to make an effort in increasing the components of $\mathbf{b}$ because it will be well compensated into a positive increment of the value of M? Which component is the best to do this? Are there some components of $\mathbf{b}$ upon which the value of M does not depend? All of these rather interesting questions can be summarized mathematically, and responded in practice, once we know the gradient $\nabla M(\mathbf{b})$. This piece of information is what we are asking for.

When computing the gradient of an arbitrary function $f(\mathbf{x})$ of several variables $\mathbf{x} = (x_1, x_2, \ldots, x_N)$, we need to examine the incremental quotients

$$\frac{f(\mathbf{x} + h\mathbf{e}_i) - f(\mathbf{x})}{h}$$

for $h \in \mathbb{R}$, small, and every $i \in \{1, 2, \ldots, N\}$. Here $\{\mathbf{e}_k\}$ is the canonical basis of $\mathbb{R}^N$. In our particular situation, when $M(\mathbf{b})$ is defined through a linear programming problem, we immediately find an unexpected difficulty: if we change slightly one component of $\mathbf{b}$, the feasible set of the linear programming problem changes! It looks as if the ground would move under our feet! Though one could try to understand how that feasible set would change with changes in $\mathbf{b}$, it looks like a more promising strategy to come up with an equivalent formulation of the problem in such a way that vector $\mathbf{b}$ does not participate in the feasible set, and in this way, it does not change with changes in $\mathbf{b}$. Is this possible? Fortunately, the answer is yes, and it is one of the most appealing features of linear programming.

One of the best ways to motivate the dual problem and insist that it is like a different side of the same coin, relies on optimality conditions that should be the same for both interpretations of the problem: the primal, and the dual. Here we refer to optimality conditions for programming problems as will be discussed in the next chapter, and so we anticipate those for linear programming problems. Optimality conditions are like special properties that optimal solutions enjoy precisely because of the fact that they are optimal for a particular problem.

Proposition 2.2 *Let $\bar{x}$ be optimal for (2.3). Then there are vectors $\bar{y} \in \mathbb{R}^n$, $\bar{z} \in \mathbb{R}^N$ such that*

$$c - \bar{y}A + \bar{z} = 0,$$
$$\bar{y} \cdot (A\bar{x} - b) = 0, \quad \bar{z} \cdot \bar{x} = 0,$$
$$\bar{y} \geq 0, \quad \bar{z} \geq 0.$$

We will come back to this proposition within the next chapter.

The first condition in the conclusion of Proposition 2.2 may be used to eliminate $\bar{z}$ from the others. If we do so, we are left with

$$\bar{y} \cdot (A\bar{x} - b) = 0, \quad (\bar{y}A - c) \cdot \bar{x} = 0,$$
$$\bar{y} \geq 0, \quad \bar{y}A - c \geq 0.$$

But recall that we also know that

$$A\bar{x} - b \geq 0, \quad \bar{x} \geq 0.$$

We want to regard y as the vector variable for a new linear programming problem whose optimal solution be $\bar{y}$. According to the conditions above, constraints for this new problem will come in the form

$$yA - c \geq 0, \quad y \geq 0.$$

To deduce the (linear) cost functional for the new problem, suppose that both y, and x are feasible for the first, and the new problems, respectively,

$$yA - c \geq 0, \quad y \geq 0,$$
$$Ax - b \geq 0, \quad x \geq 0.$$

Paying attention to the signs of the various terms involved in the inner products, it is elementary to argue that

$$c \cdot x \leq yAx \leq y \cdot b.$$

Given that y, and x are arbitrary, feasible vectors for their respective problems, we clearly see that the cost for the new (dual) problem should be $y \cdot b$, and we ought to look for the minimum instead of the maximum.

Definition 2.2 The two linear programming problems

$$\text{Maximize in } \mathbf{x}: \quad \mathbf{c} \cdot \mathbf{x} \quad \text{subject to} \quad A\mathbf{x} \leq \mathbf{b}, \mathbf{x} \geq \mathbf{0},$$

and

$$\text{Minimize in } \mathbf{y}: \quad \mathbf{b} \cdot \mathbf{y} \quad \text{subject to} \quad \mathbf{y}A \geq \mathbf{c}, \mathbf{y} \geq \mathbf{0},$$

are dual problems of each other.

There is a pretty interesting interpretation of the dual problem from a practical viewpoint which is worth to bear in mind. The dual variables $\mathbf{y}$ can be interpreted as prices of resources (sometimes called shadow prices) so that in the dual problem we would like to minimize the amount of money paid for resources provided those do not lead to cheaper prices for products. Seen those two problems from this perspective, it is enlightening to realize how they are two sides of the same coin.

Probably, the neatest way to express the intimate relationship between these two problems, and, in particular, between the optimal solutions of both is contained in the next statement.

Proposition 2.3 *Two vectors $\bar{\mathbf{x}}$, and $\bar{\mathbf{y}}$ are optimal solutions of the primal and dual problems, respectively, as above, if and only if*

$$\bar{\mathbf{y}} \cdot (A\bar{\mathbf{x}} - \mathbf{b}) = 0, \quad (\bar{\mathbf{y}}A - \mathbf{c}) \cdot \bar{\mathbf{x}} = 0,$$
$$\bar{\mathbf{y}}A - \mathbf{c} \geq \mathbf{0}, \quad \bar{\mathbf{y}} \geq \mathbf{0},$$
$$A\bar{\mathbf{x}} - \mathbf{b} \leq \mathbf{0}, \quad \bar{\mathbf{x}} \geq \mathbf{0}.$$

There are some relevant consequences from these conditions worth explicitly stating.

1. It is immediate that $\bar{\mathbf{y}} \cdot \mathbf{b} = \mathbf{c} \cdot \bar{\mathbf{x}}$, and so we conclude that the optimal value for both problems (the maximum for the primal, and the minimum for the dual) is the same number.
2. Even though the conditions

$$\bar{\mathbf{y}} \cdot (A\bar{\mathbf{x}} - \mathbf{b}) = 0, \quad (\bar{\mathbf{y}}A - \mathbf{c}) \cdot \bar{\mathbf{x}} = 0,$$

may look like just as a couple of identities, they are much more because of the sign conditions coming from feasibility. Let us focus on the first one, as the second is shown in exactly the same way. Note that

$$\overline{\mathbf{y}} \cdot (\mathbf{A}\overline{\mathbf{x}} - \mathbf{b}) = 0, \quad \mathbf{A}\overline{\mathbf{x}} - \mathbf{b} \le \mathbf{0}, \overline{\mathbf{y}} \ge \mathbf{0}.$$

The sign conditions imply that each term in the inner product $\overline{\mathbf{y}} \cdot (\mathbf{A}\overline{\mathbf{x}} - \mathbf{b})$ is non-positive because it is the product of two numbers of different signs (or zero). How can a sum of non-positive numbers add up to zero, if there is no possibility of cancelation precisely because the sign of all of them is the same? The only possibility is that every single term vanishes $\overline{y}_k((\mathbf{A}\overline{\mathbf{x}})_k - b_k) = 0$ for all k. Similarly, $\overline{x}_j((\overline{\mathbf{y}}\mathbf{A})_j - c_j) = 0$, for all j.

At this stage we are ready to provide an explicit answer to the initial issue which motivated this analysis, and the introduction of the dual problem. Recall that $M(\mathbf{b})$ is the maximum value of the cost $\mathbf{c} \cdot \mathbf{x}$ among feasible vectors $\mathbf{x}$, i.e., if $\overline{\mathbf{x}}$ is an optimal solution of the primal problem, then $M(\mathbf{b}) = \mathbf{c} \cdot \overline{\mathbf{x}}$. We now know that $M(\mathbf{b}) = \mathbf{b} \cdot \overline{\mathbf{y}}$ if $\overline{\mathbf{y}}$ is an optimal solution for the dual. But, since the feasible set for the dual, $\mathbf{y}\mathbf{A} \ge \mathbf{c}$, $\mathbf{y} \ge \mathbf{0}$, does not depend in any way on vector $\mathbf{b}$, we can conclude that $\nabla M(\mathbf{b}) = \overline{\mathbf{y}}$, the optimal solution of the dual.

> **Definition 2.3** The optimal solution $\overline{\mathbf{y}}$ of the dual problem is called the vector of sensitivity parameters of the problem. Each of its components furnishes a measure of how the value of the maximum of the primal varies with changes in that particular component for the vector of resources $\mathbf{b}$.

2.4 A Final Clarifying Example

Given a primal problem like (2.3), dual variables $\mathbf{y}$ are associated, in a one-to-one manner, with constrains in the primal: each y_k corresponds to the kth constraint $(\mathbf{A}\mathbf{x})_k - b_k \le 0$. Vice versa, each primal variable x_j corresponds to the jth constraint in the dual $(\mathbf{y}\mathbf{A})_j - c_j \ge 0$. Optimality is expressed, then, by demanding that the corresponding products variable $\times$ constraint vanish:

$$\overline{y}_k((\mathbf{A}\overline{\mathbf{x}})_k - b_k) = 0, \quad \overline{x}_j((\overline{\mathbf{y}}\mathbf{A})_j - c_j) = 0,$$

for all k, and j, in addition to the sign restrictions.

To see how these important conditions can be used in practice, we are going to work out the following explicit example.

Example 2.5 A farmer uses three types of milk, sheep, goat, and cow, to produce five different dairy products identified as Vi, $i = 1, 2, 3, 4, 5$. The next table sums up all of the relevant information about the proportions of each type of milk for a unity of each product Vi, the market price of each product, and the total resources, for a given period of time, of each type of milk.

Product	V1	V2	V3	V4	V5	Resources
Sheep	4	10	8	10	5	6000
Goat	2	1	6	30	3	4000
Cow	3	2	5	15	5	5000
Price	12	20	18	40	10	

With all of this information, it is not difficult to write down the optimization problem to find the optimal (maximizing benefits) production of the five dairy products. Namely, if x_i, $i = 1, 2, 3, 4, 5$, indicates the number of units to be produced of each Vi, then we seek to maximize

$$12x_1 + 20x_2 + 18x_3 + 40x_4 + 10x_5$$

restricted to

$$4x_1 + 10x_2 + 8x_3 + 10x_4 + 5x_5 \leq 6000,$$
$$2x_1 + x_2 + 6x_3 + 30x_4 + 3x_5 \leq 4000,$$
$$3x_1 + 2x_2 + 5x_3 + 15x_4 + 5x_5 \leq 5000,$$
$$x_1, x_2, x_3, x_4, x_5 \geq 0.$$

According to the rule to find the dual problem, proceeding with a bit of care, we find that for the dual we should minimize

$$6000y_1 + 4000y_2 + 5000y_3$$

subject to

$$4y_1 + 2y_2 + 3y_3 \geq 12,$$
$$10y_1 + y_2 + 2y_3 \geq 20,$$
$$8y_1 + 6y_2 + 5y_3 \geq 18,$$
$$10y_1 + 30y_2 + 15y_3 \geq 40,$$
$$5y_1 + 3y_2 + 5y_3 \geq 10,$$
$$y_1, y_2, y_3 \geq 0.$$

The whole point of the relationship between these two problems is expressed in the specific and intimate link between constraints and variables in the following way

$$4x_1 + 10x_2 + 8x_3 + 10x_4 + 5x_5 \leq 6000 \implies y_1 \geq 0,$$
$$2x_1 + x_2 + 6x_3 + 30x_4 + 3x_5 \leq 4000 \implies y_2 \geq 0,$$
$$3x_1 + 2x_2 + 5x_3 + 15x_4 + 5x_5 \leq 5000 \implies y_3 \geq 0,$$
$$4y_1 + 2y_2 + 3y_3 \geq 12 \implies x_1 \geq 0,$$
$$10y_1 + y_2 + 2y_3 \geq 20 \implies x_2 \geq 0,$$
$$8y_1 + 6y_2 + 5y_3 \geq 18 \implies x_3 \geq 0,$$
$$10y_1 + 30y_2 + 15y_3 \geq 40 \implies x_4 \geq 0,$$
$$5y_1 + 3y_2 + 5y_3 \geq 10 \implies x_5 \geq 0,$$

and at the level of optimality, for optimal solutions of both, we should have

$$y_1(4x_1 + 10x_2 + 8x_3 + 10x_4 + 5x_5 - 6000) = 0,$$
$$y_2(2x_1 + x_2 + 6x_3 + 30x_4 + 3x_5 - 4000) = 0,$$
$$y_3(3x_1 + 2x_2 + 5x_3 + 15x_4 + 5x_5 - 5000) = 0,$$
$$x_1(4y_1 + 2y_2 + 3y_3 - 12) = 0,$$
$$x_2(10y_1 + y_2 + 2y_3 - 20) = 0,$$
$$x_3(8y_1 + 6y_2 + 5y_3 - 18) = 0,$$
$$x_4(10y_1 + 30y_2 + 15y_3 - 40) = 0,$$
$$x_5(5y_1 + 3y_2 + 5y_3 - 10) = 0.$$

Assume we are given the following three pieces of information:

1. the maximum possible benefit is 18,400;
2. the optimal solution for the primal is in need of $V1$;
3. for the optimal solution the total amount of cow milk is not fully spent.

Let us interpret all of this information, and see if we can deduce the optimal solutions for both problems. There is nothing to say about the maximum value, except that is also the optimal value for the minimum in the dual

$$18400 = 6000y_1 + 4000y_2 + 5000y_3 = 12x_1 + 20x_2 + 18x_3 + 40x_4 + 10x_5.$$

The second piece of information is telling us that $x_1 > 0$ because there is no way to find the optimal solution if we insist in being dispensed with $V1$. But then the restriction corresponding to x_1 in the above list should vanish

$$4y_1 + 2y_2 + 3y_3 = 12.$$

Finally, the restriction for the cow milk is strict, and hence the corresponding dual variable must vanish $y_3 = 0$.

Altogether, we have

$$18400 = 6000y_1 + 4000y_2 + 5000y_3, \quad 4y_1 + 2y_2 + 3y_3 = 12, \quad y_3 = 0.$$

It is easy to find the optimal solution of the dual $(14/5, 2/5, 0)$. If we take back these values of the dual variables to all of the above products, we immediately find that $x_2 = x_3 = x_5 = 0$, and

$$4x_1 + 10x_4 = 6000, \quad 2x_1 + 30x_4 = 4000,$$

i.e. $x_1 = 1400, x_4 = 40$, and the optimal solution for the primal is $(1400, 0, 0, 40, 0)$.

We now know the optimal production to maximize benefits. As we have insisted above the solution of the dual problem $(14/5, 2/5, 0)$ are the sensitivity parameters expressing how sensitive the value $18,400$ of the maximum is with respect to the three resources constraints. In particular, it says that it is unsensitive to the cow milk constraint, and so there is no point in trying to increase the amount of cow milk at our disposal (this is also seen in the fact that there is a certain surplus of cow milk as we were informed before). However, we can expect an increment of about $2/5$ monetary units in the benefit for each unity we increase the quantity of goat milk at our disposal, and around $14/5$ units, if we do so for sheep milk.

2.5 Final Remarks

Linear programming looks easy, and indeed understanding its structure is not that hard, if one feels comfortable with the main concepts of Linear Algebra. However, even with linear programming, realistic problems can be very hard to treat mainly because of its size in terms of number of variables, and number of constraints involved. Even for moderate sized problems, it is impossible to find solutions by hand. So, in practice, there is no substitute for a good computational package to find accurate approximation of optimal solutions. We will treat this fundamental issue in the final chapter of this part of the book.

But there is more to linear programming. More often than not, feasible vectors for problems cannot take on non-integer values, so that quantities can only be measured in full units. That is the case of tangible objects (chairs, tables, cars, etc.), or non-tangible ones like full trips, full steps, etc. What we mean is that in these cases, as part of feasibility, variables can only take on integer, or even non-negative integer values. In this case, we talk about integer programming. This apparent innocent condition puts many complex difficulties to the problem. We can always ignore those integer-valued constrains, and cut decimal figures at the end, and this indeed may provide very good solutions, but they may not be the best ones. Some times variables can only take two values: 0, and 1, as a way to model a two-possibility decision, and in this case we talk about binary variables.

The treatment of duality and sensitivity is not complete either. A thorough exam-
ination of special situations when one of the two versions, the primal or the dual,
turns out to be infeasible or not admit an optimal solution; the possibility of having
a gap between the optimal values of the two problems, etc., would require additional
effort.

There are also important and difficult issues associated with large scale problems,
and how to divide them up into pieces so that the solution procedure is as cheap (in
time, in money) as possible. Many other areas of linear programming are reserved
to experts.

Despite the simplicity of the structure of LP problems, its applicability is per-
vasive throughout science and engineering: agriculture, economics, management,
manufacturing, telecommunication, marketing, finance, energy,...

2.6 Exercises

2.6.1 Exercises to Support the Main Concepts

1. Find the extreme values of the function

$$f(x_1, x_2, x_3) = x_1 - x_2 + 5x_3$$

in the convex set (polyhedron) generated by the points (vertices)

$$(1, 1, 0), \quad (-1, 0, 1), \quad (2, 1, -1), \quad (-1, -1, -1).$$

Same question for the function

$$g(x_1, x_2, x_3) = x_2 - x_3.$$

2. Consider the LP problem

$$\text{Maximize in } (x_1, x_2): \quad x_1 + 3x_2$$

under

$$-x_1 + x_2 \leq 2, \quad 4x_1 - x_2 \leq -1, \quad 2x_1 + x_2 \geq 0, \quad x_1, x_2 \geq 0.$$

(a) Find its optimal solution graphically.
(b) Solve the three problems consisting in maximizing the same cost in the three
different feasible sets

$$-x_1 + x_2 \leq 2 + h, \quad 4x_1 - x_2 \leq -1, \quad 2x_1 + x_2 \geq 0,$$
$$-x_1 + x_2 \leq 2, \quad 4x_1 - x_2 \leq -1 + h, \quad 2x_1 + x_2 \geq 0,$$
$$-x_1 + x_2 \leq 2, \quad 4x_1 - x_2 \leq -1, \quad 2x_1 + x_2 \geq h,$$

where h is a real (positive or negative) parameter. Note how the feasible set changes with h in each situation.

(c) Calculate the derivatives of the optimal value with respect to h for these three cases.

(d) Solve the dual problem of the initial one, and relate its optimal solution to the previous item.

3. Some times LP problems may not naturally come in the format we have been using. But with some extra effort, they can be given that form, and then find the dual. Look at the problem

$$\text{Maximize in } (x_1, x_2): \quad x_1 + 3x_2$$

under

$$-x_1 + x_2 \leq 2, \quad 4x_1 - x_2 \leq -1, \quad 2x_1 + x_2 \geq 0,$$

but with no constraint on the signs of x_1 and x_2. Make the change of variables

$$X_1 = 2 + x_1 - x_2, \quad X_2 = -1 - 4x_1 + x_2,$$

and write the problem in terms of (X_1, X_2). Find the dual. Try other possible changes of variables.

4. For the LP problem

$$\text{Maximize in } \mathbf{x}: \quad \mathbf{c} \cdot \mathbf{x} \quad \text{under} \quad A\mathbf{x} \leq \mathbf{b},$$

write each variable x_i as the difference of two non-negative variables x_{i+}, x_{i-} (its positive and negative parts, respectively) to check that the dual problem is

$$\text{Minimize in } \mathbf{y}: \quad \mathbf{b} \cdot \mathbf{y} \quad \text{under} \quad \mathbf{y}A = \mathbf{c}, \mathbf{y} \geq 0.$$

5. Find the optimal solution of the problem

$$\text{Maximize in } (x_1, x_2, x_3): \quad x_1 + x_2 - x_3$$

subject to

$$x_1 - x_2 + x_3 \leq 1, \quad x_1 + x_2 + x_3 \leq 1, \quad x_1, x_2, x_3 \geq 0.$$

6. Let's have a look at the problem

$$\text{Maximize in } (x_1, x_2, x_3, x_4): \quad x_1 + x_2 - x_3 - x_4$$

subject to

$$x_1 - x_2 + x_3 - x_4 \leq 1, \quad x_1 - x_2 - x_3 + x_4 \leq 1,$$
$$x_1, x_2, x_3, x_4 \geq 0.$$

(a) Write the dual problem.
(b) Argue that the dual problem is infeasible.
(c) Given that the maximum for the primal is the infimum for the dual, conclude that the primal problem cannot have a solution.
(d) By taking $x_3 = x_4 = 0$, reduce the primal to a problem in dimension two. Check graphically that for this subproblem the maximum is infinite.

7. Find the optimal solution of the problem

$$\text{Minimize in } (x_1, x_2): \quad 2x_1 - 7x_2$$

under

$$x_1 + hx_2 \leq 0, \quad hx_1 - 2x_2 \geq -1, \quad x_1, x_2 \geq 0,$$

in terms of the parameter $h < 0$.

8. For the LP problem

$$\text{Maximize in } (x_1, x_2, x_3): \quad x_1 + 5x_2 - 3x_3$$

under

$$-x_1 + x_2 + x_3 \leq 1, \quad x_1 - x_2 + x_3 \leq 1, \quad x_1 + x_2 - x_3 \leq 1,$$
$$x_1, x_2, x_3 \geq 0,$$

find its dual. Transforming this dual problem in a form like our main model problem in this chapter, find its dual and compare it to the original problem.

9. Find the maximum of the cost function $18x_1 + 4x_2 + 6x_3$ over the set

$$3x_1 + x_2 \leq -3, \quad 2x_1 + x_3 \leq -5, \quad x_1, x_2, x_3 \leq 0,$$

through its dual.

10. Decide for which values of k the LP problem

$$\text{Optimize in } (x_1, x_2) \in \mathbf{F}: \quad x_1 + kx_2,$$

where

$$F = \{(x_1, x_2) \in \mathbb{R}^2 : 3x_1 + 2x_2 \geq 6, \ x_1 + 6x_2 \geq 8, \ x_1, x_2 \geq 0\},$$

has an optimal solution. Discuss when such a solution is unique (optimize means both minimize and maximize).

2.6.2 Practice Exercises

1. Draw the region limited by the conditions

$$x_2 \geq 0, \quad 0 \leq x_1 \leq 3, \quad -x_1 + x_2 \leq 1, \quad x_1 + x_2 \leq 4.$$

For each one of the following objective functions, find the points where the maximum is achieved

$$f_1 = 2x_1 + x_2, \quad f_2 = x_1 + x_2, \quad f_3 = x_1 + 2x_2.$$

2. Solve graphically the following problems

Maximize	$2x_1 + 6x_2$	Maximize	$-3x_1 + 2x_2$
under	$-x_1 + x_2 \leq 1,$	under	$x_1 + x_2 \leq 5,$
	$2x_1 + x_2 \leq 2,$		$0 \leq x_1 \leq 4,$
	$x_1, x_2 \geq 0,$		$1 \leq x_2 \leq 6.$

3. Find the maximum of the function $240x_1 + 104x_2 + 60x_3 + 19x_4$ over the set determined by the constraints

$$20x_1 + 9x_2 + 6x_3 + x_4 \leq 20,$$
$$10x_1 + 4x_2 + 2x_3 + x_4 \leq 10,$$
$$x_1, x_2, x_3, x_4 \geq 0.$$

4. For the problem

$$\text{Minimize} \quad 20x_1 - 10x_2 - 6x_3 + 2x_4$$
$$\text{under} \ \ 2x_1 + 8x_2 + 6x_3 + x_4 \leq 20$$
$$10x_1 + 4x_2 + 2x_3 + x_4 \leq 10$$
$$x_1 - 5x_2 + 3x_3 + 8x_4 \geq 1$$
$$x_i \geq 0, \ i = 1, 2, 3, 4,$$

someone has provided the following information:

(a) the minimum value is -22;

(b) in the optimal solution, the second constraint is an strict inequality;
(c) the third restriction of the corresponding dual problem over its optimal solution is an equality.

Write the dual problem, and find the optimal solutions of both.

5. Solve the problem

$$\max_{(x_1,x_2,x_3)\in F} 2x_1 + 3x_2 + x_3$$

where F is the subset of $\mathbb{R}^3$ defined by

$$x_1 + x_2 + 2x_3 \le 200, \quad 2x_1 + 2x_2 + 5x_3 \le 500, \quad x_1 + 2x_2 + 3x_3 \le 300,$$
$$x_1, x_2, x_3 \ge 0.$$

6. Find the optimal solution of Example 2.2. Which one of the four participating cables would allow for a greater increase of the overall weight the system can withstand if resistance is increase in a unit?

2.6.3 Case Studies

1. A certain company makes a unit of four different jobs A, B, C, and D, with a specified amount of labor O and resources P according to the table below. Moreover, there is a given number of labor units and resources at its disposal. Market prices are also known.

	Job A	Job B	Job C	Job D	Availability
Labor O	1	2	2	4	200
Resources P	2	1	3	1	150
Price	20	30	30	50	

(a) Formulate the problem of maximizing benefits. Write its dual.
(b) Find the optimal solutions for both, and the maximum benefit possible.
(c) Interpret the solution of the dual in terms of the primal. What is more advantageous: to increase labor, or resources?

2. In a carpenter's shop, a clerk may decide to work on tables, chairs or stools. For the first, he receives 3.35 euros per unit; 2, for the each chair; and 0.5 euros for the last. Tables require 1.25 h each; chairs, 0.75 h; and stools, 0.25 h. If each workday amounts to 8 h, and each unfinished element is worthless, decide how many pieces of each kind this clerk has to make in order to maximize salary.

3. A thief, after committing a robbery in a jewelry, hides in a park which is the convex hull of the four points

$$(0, 0), \quad (-1, 1), \quad (1, 3), \quad (2, 1).$$

The police, looking after him, gets into the park and organizes the search according to the function

$$p(x_1, x_2) = x_1 - 3x_2 + 10$$

indicating density of vigilance. Recommend the thief the best point through which he can escape, or the best point where he can stay hidden.

4. A farmer wants to customize his fertilizer for his current crop.[1] He can buy plant food mix A and plant food mix B. Each cubic yard of food A contains 20 lb of phosphoric acid, 30 lb of nitrogen and 5 lb of potash. Each cubic yard of food B contains 10 lb of phosphoric acid, 30 lb of nitrogen and 10 lb of potash. He requires a minimum of 460 lb of phosphoric acid, 960 lb of nitrogen and 220 lb of potash. If food A costs 30 per cubic yard and food B costs 35 per cubic yard, how many cubic yards of each food should the farmer blend to meet the minimum chemical requirements at a minimal cost? What is this cost?

5. A company makes three models of desks[2]: an executive model, an office model and a student model. Each desk spends time in the cabinet shop, the finishing shop and the crating shop as shown in the table:

	Cabinet	Finishing	Crating	Profit
Executive	2	1	1	150
Office	1	2	1	125
Student	1	1	0.5	50
Hours	16	16	10	

How many of each type of model should be made to maximize profits?

6. A small brewery produces ale and beer from corn, hops and barley malt, and sells the product according to the data in the following table[3]:

	Corn	Hops	Malt	Profit
Available	480	160	1190	
Ale	5	4	35	13
Beer	15	4	20	23

Choose product mix to maximize profits.

7. A calculator company produces a scientific calculator and a graphing calculator.[4] Long-term projections indicate an expected demand of at least 100 scientific and

---

[1] From http://www.math.tamu.edu/~Janice.Epstein/141/review/LinProgReview.pdf.

[2] From http://www.math.tamu.edu/~Janice.Epstein/141/review/LinProgReview.pdf.

[3] From https://www.cs.princeton.edu/~rs/AlgsDS07/22LinearProgramming.pdf.

[4] From http://www.purplemath.com.

80 graphing calculators each day. Because of limitations on production capacity, no more than 200 scientific and 170 graphing calculators can be made daily. To satisfy a shipping contract, a total of at least 200 calculators much be shipped each day. If each scientific calculator sold results in a 2 loss, but each graphing calculator produces a 5 profit, how many of each type should be made daily to maximize net profits?

8. You need to buy some filing cabinets.[5] You know that Cabinet X costs 10 per unit, requires six square feet of floor space, and holds eight cubic feet of files. Cabinet Y costs 20 per unit, requires eight square feet of floor space, and holds twelve cubic feet of files. You have been given 140 for this purchase, though you don't have to spend that much. The office has room for no more than 72 square feet of cabinets. How many of which model should you buy, in order to maximize storage volume?

9. A manufacturer produces violins, guitars, and viola from quality wood, metal and labor according to the table

	Violin	Guitar	Viola
Wood	2	1	1
Labor	2	1	2
Metal	1	1	1

Disposal is 50 units of wood, 60 labor units and 55 units of metal; prices are: 200 for violins, 175 for guitars, and 125 for violas.

(a) Formulate the problem and find the optimal production plan.
(b) Write its dual, and find its optimal solution through the optimal solution of the primal problem.
(c) How much more would the company be willing to pay for an extra unit of wood? For an extra unit of labor? And for one of metal?
(d) What is the range of prices to which we could sell guitars without compromising the optimal production plan found earlier?

---

[5]From http://www.purplemath.com.

Chapter 3
Nonlinear Programming

Non-linear programming problems are basically just as their linear counterparts in the sense that their formal structure is similar, with the difference that some non-linear functions occur. This sentence is definitely misleading as it may give the (false) impression that understanding non-linear optimization problems is more-or-less a generalization of the linear problems. Nothing farther from reality, as we will immediately understand even with the simplest non-linear problems. This chapter focuses on two main fundamental topics: optimality conditions, and convexity.

3.1 Some Motivating Examples

We have selected three interesting problems.

Example 3.1 The building problem (taken directly from [5]). We would like to design a building in the form of a big cube with the objective in mind to limit costs for heating and cooling, and save some money in the construction process. Refer to Fig. 3.1. The building can be partially underground, and there is a number of conditions to be respected. The various variables involved are:

- n is the number of floors, a positive integer;
- d is the depth of the building below ground;
- h is the height of the building above ground;
- l is the length, and w is the width.

The conditions to be respected are:

- the global floor space required is at least 20000 m^2;
- the lot dimensions require $w, l \leq 50$;
- for aesthetical purposes, l/w must be the golden ratio $r = 1.618$;
- each floor must be 3.5 m high;

© Springer International Publishing AG 2017
P. Pedregal, *Optimization and Approximation*, UNITEXT 108,
DOI 10.1007/978-3-319-64843-9_3

Fig. 3.1 The building of
Example 3.1

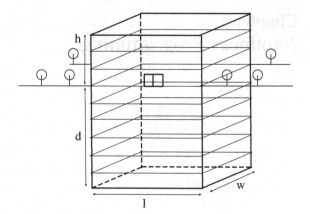

- heating and cooling costs are estimated at 100 monetary units per m^2 for exposed surface of the building;
- total heating and cooling costs, on a annual basis, cannot exceed 225000 monetary units.

The objective is to minimize the volume that has to be excavated.

Let us try to translate all of the information in equations. The objective is to minimize the function dlw, and the restrictions to be respected can be written down in the form

$$nlw \geq 20000, \quad l = rw \leq 50, \quad d + h = 3.5n,$$
$$100(2hl + 2hw + lw) \leq 225000.$$

We neglect the fact that n should be an integer. We can use the two identities

$$l = rw, \quad d + h = 3.5n$$

to eliminate the two variables l and n from the formulation, and so simplify the statement. Namely, working out the new form of the restrictions in terms of the three variables d, w, and h, and bearing in mind that minimizing rdw^2 is equivalent to minimizing dw^2 (because r is a positive constant), we come to the final form of the problem

$$\text{Minimize in } (d, w, h): \quad dw^2$$

under

$$(d + h)w^2 \leq 70000/r, \quad w \leq 50/r, \quad 2(r + 1)hw + rw^2 \leq 2250.$$

Once the optimal solution is found, one needs to recall that $n = (d + h)/3.5$ is the number of floors (overlooking the fact that it must be an positive integer), and that $l = rw$.

Example 3.2 The investor problem (from [31] and many other places). A certain investor is thinking about dividing a given overall amount C in three different stock markets $i = 1, 2, 3$. He knows the prices p_i, and the benefits r_i for each stock, respectively.

1. If his profile is that of an aggressive investor (no fear to risk), then find the investment of maximum benefit.
2. If risk is measured through a factor $\beta > 0$, and the covariance matrix $\sigma = (\sigma_{ij})$ of each two stocks, assumed to be positive definite, the benefit is given by subtracting from the benefit of the previous item, the term $\beta x^T \sigma x$ where $x = (x_1, x_2, x_3)$ is the vector of amounts invested on each stock. Find, in this new setting, the distribution that maximizes the total benefit.

We have almost specified variables in the statement of the problem. Let $x = (x_1, x_2, x_3)$ be the distribution of the total capital C in the three stocks. Then the total benefit, if we ignore risk, will be given by the expression

$$x \cdot r = x_1 r_1 + x_2 r_2 + x_3 r_3.$$

So we want to maximize this inner product under the constraints

$$x \geq 0, \quad x \cdot p = x_1 p_1 + x_2 p_2 + x_3 p_3 \leq C.$$

This is the first part of the problem, and it corresponds to a typical linear programming situation.

For the second part, according to the statement, we would seek to

$$\text{Maximize in } x \in \mathbb{R}^3 : \quad x \cdot r - \beta x^T \sigma x$$

subject to

$$x \geq 0, \quad x \cdot p = x_1 p_1 + x_2 p_2 + x_3 p_3 \leq C.$$

The presence of the quadratic part in the cost suffices to dispatch this problem as a non-linear programming problem.

Example 3.3 A truss structure ([10]). See Fig. 3.2 for the configuration of a mechanical system. It is to be designed for a minimum total weight under the conditions that the maximum tension in nodes 1 and 2, as well as the displacement of node 3, are not to exceed certain threshold values. More specifically, the various parameters of the problem are:

Fig. 3.2 A truss structure

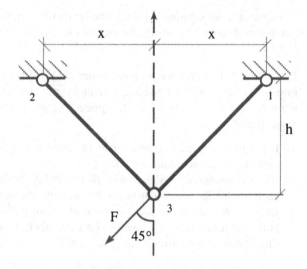

- density γ for the material the bars are made of;
- Young modulus E of the same material;
- load F on node 3, acting with a 45 degree angle to the left of vertical axis;
- maximum allowable load T_0 on nodes 1 and 2;
- maximum displacement D_0 permitted for node 3;
- height h of the structure.

The design variables are x, half of the distance between nodes 1 and 2; and z, the area of the cross section of the bars. Functions involved are:

- the total weight of the structure

$$W(x, z) = 2\gamma z \sqrt{x^2 + h^2};$$

- tension in node 1

$$T_1(x, z) = \frac{F}{2\sqrt{2}h} \frac{(h + x)\sqrt{x^2 + h^2}}{xz};$$

- tension in node 2

$$T_2(x, z) = \frac{F}{2\sqrt{2}h} \frac{(h - x)\sqrt{x^2 + h^2}}{xz};$$

- displacement of node 3

$$D(x, z) = \frac{F}{2\sqrt{2}Eh^2} \frac{(h^2 + x^2)^{3/2}(h^4 + x^4)^{1/2}}{x^2 z}.$$

The optimization problem becomes

$$\text{Minimize in } (x, z): \quad W(x, z)$$

under

$$D(x, z) \leq D_0, \quad T_1(x, z) \leq T_0, \quad T_2(x, z) \leq T_0, \quad x, z \geq 0.$$

3.2 Structure of Non-linear Programming Problems

Our model problem is

$$\text{Minimize in } \mathbf{x} \in \mathbb{R}^N: \quad f(\mathbf{x}) \quad \text{subject to} \quad \mathbf{h}(\mathbf{x}) = \mathbf{0}, \mathbf{g}(\mathbf{x}) \leq \mathbf{0}, \qquad (3.1)$$

where

$$f : \mathbb{R}^N \to \mathbb{R}, \quad \mathbf{h} : \mathbb{R}^N \to \mathbb{R}^m, \quad \mathbf{g} : \mathbb{R}^N \to \mathbb{R}^n,$$

are always assumed to be smooth. We are therefore to respect m constraints in the form of equalities, and n, in the form of inequalities.

The non-linear world hides so many surprises that one can hardly say anything in general that applies to all situations. The feasible set is

$$\mathbf{F} = \{\mathbf{x} \in \mathbb{R}^N : \mathbf{h}(\mathbf{x}) = \mathbf{0}, \mathbf{g}(\mathbf{x}) \leq \mathbf{0}\}.$$

There is no special structure for this set if we do not restrict the nature of the functions involved. The only thing one can say is that $\mathbf{F}$ is called limited or bounded, if it can be enclosed in a big ball, so that the set is limited in every possible direction; unbounded, otherwise.

As in the linear case, the concept of an optimal solution of the problem is quite clear.

Definition 3.1 A feasible vector $\overline{\mathbf{x}} \in \mathbf{F}$ is called an optimal solution of (3.1) if $f(\overline{\mathbf{x}}) \leq f(\mathbf{x})$ for all feasible $\mathbf{x} \in \mathbf{F}$.

The location of these optimal solutions is the whole concern in non-linear programming. Non-linear programs can have any number (including infinite) of optimal solutions, there is no restriction, and it is very hard to anticipate how many there will be in a given problem.

The very first issue, to avoid wasting our time, is to decide on the existence of (at least) one optimal solution. Again it is impossible to cover all possibilities, but there is definitely a general principle that is sufficient for many of the situations in practice.

Proposition 3.1 (Weierstrass) *If $f : \mathbb{R}^N \to \mathbb{R}$ is continuous, and the feasible set $\mathbf{F}$ is bounded, then the corresponding mathematical program admits optimal solutions. If it is not bounded, but still*

$$\lim_{|\mathbf{x}| \to \infty, \mathbf{x} \in \mathbf{F}} f(\mathbf{x}) = +\infty,$$

then there are optimal solutions as well.

The main task is to find them.

3.3 Optimality Conditions

Optimality conditions are fundamental. They simply indicate where to look for optimal solutions of problems, limiting a lot the places, within the feasible set, where optimal solutions could eventually occur. As such, they are necessary conditions that optimal solutions should comply with, much in the same way as points of minimum must be critical points of functions.

Let us focus on a general problem like (3.1), and suppose that $\bar{\mathbf{x}}$ is truly an optimal solution in such a way that

$$f(\bar{\mathbf{x}}) \le f(\mathbf{x}) \text{ for all } \mathbf{x} \in \mathbf{F}. \tag{3.2}$$

The admissibility of $\bar{\mathbf{x}}$ forces to have $\mathbf{h}(\bar{\mathbf{x}}) = \mathbf{0}$, but concerning inequality constraints, we can have two situations depending on whether equality or strict inequality holds. In general, there will be some components of $\mathbf{g}$ for which we will have equality $g^{(i)}(\bar{\mathbf{x}}) = 0$, and these are called active (in $\bar{\mathbf{x}}$), whereas for the other $g^{(j)}(\bar{\mathbf{x}}) < 0$, and they are inactive.

How can we exploit (3.2) to conclude something about $\bar{\mathbf{x}}$? We have to produce feasible, one-dimensional sections of f starting at $\bar{\mathbf{x}}$, and state the comparison. More specifically, take any smooth curve

$$\sigma(t) : [0, \varepsilon) \to \mathbb{R}^N, \quad \sigma(t) \in \mathbf{F}, \sigma(0) = \bar{\mathbf{x}},$$

and focus on the function of a single variable $\phi(t) = f(\sigma(t)) : [0, \varepsilon) \to \mathbb{R}$. Property (3.2) for $x = \sigma(t)$, translates into the fact that ϕ has a one-side minimum at $t = 0$, i.e. $\phi'(0) \geq 0$. By the chain rule,

$$\nabla f(\overline{x}) \cdot \sigma'(0) \geq 0. \tag{3.3}$$

We therefore see that the inner product of the gradient of f at the point of minimum $\overline{x}$ with tangent vectors of all feasible paths has to be non-negative. What tangent vectors $v = \sigma'(0)$ are feasible at $\overline{x}$? This may be, in general, a rather delicate issue which we do not pretend to examine rigorously here as we will follow a different route below. At this stage, we are contented with having a feeling about where multipliers come from.

To this end, suppose that there are just equality constraints $h = 0$ to be respected in F. Then feasible curves σ as above will have to comply with $h(\sigma(t)) \equiv 0$ for $t \in [0, \varepsilon)$. Differentiating with respect to t and letting $t = 0$ afterwards, we arrive at the restriction

$$\nabla h(\overline{x}) \cdot \sigma'(0) = 0. \tag{3.4}$$

On the other hand, because there are no inequalities involved in our mathematical program, the one-side condition (3.3) becomes this time an equality

$$\nabla f(\overline{x}) \cdot \sigma'(0) = 0.$$

This has to be true for every vector $v = \sigma'(0)$ which is in the kernel of the linear mapping with matrix $\nabla h(\overline{x})$ according to (3.4). It is a fact in Linear Algebra that this can only happen if the vector $\nabla f(\overline{x})$ belongs to the image subspace of $\nabla h(\overline{x})$, and this dependence gives rise to the multipliers y

$$\nabla f(\overline{x}) = y \nabla h(\overline{x}).$$

The case of inequalities is more involved because one has to pay attention to one-side conditions since inequality restrictions are to be enforced, and, on the other hand, some of those constraints may be active, and some other may be inactive. Beyond these considerations, there is the more subtle issue about what vectors $v = \sigma'(0)$ are allowed in (3.3). This goes directly into the topic of constraint qualifications which essentially depend on the nature of the components of the fields h and g.

Theorem 3.1 (Karush–Kuhn–Tucker optimality conditions) *Suppose that all of the functions involved in the mathematical program* (3.1) *are smooth, and let $\overline{x}$ be an optimal (local) solution for it. If the matrix with rows $\nabla h(\overline{x})$ and $\nabla g_i(\overline{x})$ for i corresponding to active constraints $g_i(\overline{x}) = 0$ is of maximal rank, then there are vectors (of multipliers) $\overline{y} \in \mathbb{R}^m, \overline{z} \in \mathbb{R}^n$, such that*

$$\nabla f(\overline{x}) + \overline{y}\nabla h(\overline{x}) + \overline{z}\nabla g(\overline{x}) = 0,$$
$$\overline{z} \cdot g(\overline{x}) = 0,$$
$$h(\overline{x}) = 0,$$

and, in addition,

$$\overline{z} \geq 0, \quad g(\overline{x}) \leq 0.$$

Before discussing the proof of this important result, there are some relevant remarks to be made about these conditions.

1. This statement points to a certain set of vectors, those that respect all optimality conditions, where we can find the solutions of our problem. The interest of such conditions is that we can use them to find (or try to find) the potential candidates for minimizers. In this way, we simply write x, y, z, (instead of $\overline{x}$, $\overline{y}$, $\overline{z}$), and look for those triplets (x, y, z) that comply with

$$\nabla f(x) + y\nabla h(x) + z\nabla g(x) = 0,$$
$$z \cdot g(x) = 0,$$
$$h(x) = 0,$$
$$z \geq 0, \quad g(x) \leq 0.$$

2. We have separated those conditions in the form of equalities, and those in the form of inequalities, because in practice, when applying them to specific problems, they play a different role. All of the equalities form a big non-linear system that can have various solutions, among which we are to find minimizers. Inequalities are then used to sort them out, discarding those candidates which do not respect all those inequalities.
3. Optimality conditions in the statement correspond to the minimum. For the maximum, there are some changes with minus signs that need to be performed with extra care.
4. At first sight, when counting equations and unknowns x, y, z, it may look as if we are running short of equations. Indeed, the condition $z \cdot g(x) = 0$ looks like a single equation, but it is not so. Notice that if $z \geq 0$, $g(x) \leq 0$, and $z \cdot g(x) = 0$, the only possibility is that $z^{(i)}g^{(i)}(x) = 0$, independently for all components i.
5. Solving a non-linear system might be a nightmare most of the time. It may not be clear even where to start. However, the structure of product equations $z^{(i)}g^{(i)}(x) = 0$, suggest at least one way to organize calculations. Each such equation is correct when either $z^{(i)} = 0$, or $g^{(i)}(x) = 0$, and so for each i, we have two possibilities. Altogether, we can proceed to solve the system, by discussing 2^n different cases. Even for moderate values of n, this is unfeasible. However, it is also true that at least in simple examples, many of those 2^n possibilities yield no solution.

Proof (of Theorem 3.1) We proceed in two successive steps. Suppose first that we only have equality constraints to preserve so that we face the program

Minimize in $\mathbf{x}$: $f(\mathbf{x})$ subject to $\mathbf{h}(\mathbf{x}) = \mathbf{0}$.

Let us also denote by $\mathbf{x}$ the point where the minimum is attained. By our hypothesis on the linear independence of the rows of the matrix $\nabla \mathbf{h}(\mathbf{x})$, we conclude that for every vector $\mathbf{v}$ in the kernel of $\nabla \mathbf{h}(\mathbf{x})$, we can find a curve $\sigma(t) : (-\varepsilon, \varepsilon) \to \mathbb{R}^N$ whose full image is contained in the feasible set $\mathbf{h}(\sigma(t)) = \mathbf{0}$ for all such t's, and $\sigma(0) = \mathbf{x}$. Technically, we could say that the set of vectors $\mathbf{h} = \mathbf{0}$, under the full-rank condition, determines a smooth manifold whose tangent vector space at $\mathbf{x}$ is precisely the set of tangent vectors of curves like σ, and it can be identified with the kernel of $\nabla \mathbf{h}(\mathbf{x})$. The composition $f(\sigma(t)) : (-\varepsilon, \varepsilon) \to \mathbb{R}$ is smooth, and has a minimum at $t = 0$. Hence, its derivative at that value ought to vanish

$$0 = \left. \frac{d}{dt} f(\sigma(t)) \right|_{t=0} = \nabla f(\sigma(0)) \cdot \sigma'(0) = \nabla f(\mathbf{x}) \cdot \mathbf{v}.$$

The arbitrariness of $\mathbf{v} \in \ker \nabla \mathbf{h}(\mathbf{x})$ clearly implies that $\nabla f(\mathbf{x})$ belongs to the image subspace of $\nabla \mathbf{h}(\mathbf{x})$, and this dependence gives rise to the multipliers

$$\nabla f(\mathbf{x}) + \mathbf{y} \nabla \mathbf{h}(\mathbf{x}) = \mathbf{0}.$$

This has already been explained earlier.

For the second step, where we would like to consider constraints both in the form of equalities and inequalities, we will be using the following interesting lemma.

Lemma 3.1 *Let $G(\mathbf{u}) : \mathbb{R}^n \to \mathbb{R}$ attain its minimum at $\mathbf{u}_0$ among all those vectors $\mathbf{u}$ with non-positive components $\mathbf{u} \leq \mathbf{0}$. If G is smooth at $\mathbf{u}_0$, then*

$$\nabla G(\mathbf{u}_0) \cdot \mathbf{u}_0 = 0, \quad \nabla G(\mathbf{u}_0) \leq \mathbf{0}, \mathbf{u}_0 \leq \mathbf{0}.$$

Proof The lemma is indeed elementary. It refers to one-sided minima. Note that for a function $g(u)$ of a single variable, we would have two possibilities: either $g'(u_0) = 0$ (with $u_0 \leq 0$), or else $u_0 = 0$, and then $g'(0) \leq 0$. At any rate, the conclusion in the statement of the lemma is correct. The same argument can be easily extended to several variables. Indeed, if $\mathbf{u}_0$ is the point of minimum of G, for those vanishing components of $\mathbf{u}_0$, we would have that the corresponding partial derivative of G cannot be strictly positive, while for those non-null components, the partial derivative must vanish. This is exactly the conclusion in the statement.

We go back to the non-linear, general program (3.1). Take $\mathbf{u} \in \mathbb{R}^n$ with $\mathbf{u} \leq \mathbf{0}$, and consider the problem

$$\text{Minimize in } \mathbf{x} : \quad f(\mathbf{x}) \quad \text{subject to} \quad \mathbf{h}(\mathbf{x}) = \mathbf{0}, \mathbf{g}(\mathbf{x}) = \mathbf{u}. \qquad (3.5)$$

Let us designate by $G(\mathbf{u})$ the value of this minimum as a function of $\mathbf{u}$. It is clear that G attains its minimum at $\overline{\mathbf{u}} = \mathbf{g}(\overline{\mathbf{x}})$. By our assumption on the linear independence of the various gradients involved, as in the statement of the theorem, we can conclude that

G is differentiable at $\overline{\mathbf{u}}$. Even more, in a suitable vicinity of $\overline{\mathbf{u}}$, the mapping $\mathbf{u} \mapsto \mathbf{x}(\mathbf{u})$ taking each $\mathbf{u}$ into the point of minimum $\mathbf{x}(\mathbf{u})$ for (3.5) is also differentiable. By our previous step when we only permitted equality constraints as in (3.5), we could find vectors $(\mathbf{y}, \mathbf{z}) \in \mathbb{R}^m \times \mathbb{R}^n$ such that

$$\nabla f(\mathbf{x}(\mathbf{u})) + \mathbf{y}(\mathbf{u})\nabla \mathbf{h}(\mathbf{x}(\mathbf{u})) + \mathbf{z}(\mathbf{u})\nabla \mathbf{g}(\mathbf{x}(\mathbf{u})) = \mathbf{0}, \tag{3.6}$$

$$\mathbf{h}(\mathbf{x}(\mathbf{u})) = \mathbf{0}, \quad \mathbf{g}(\mathbf{x}(\mathbf{u})) = \mathbf{u}. \tag{3.7}$$

The mappings $\mathbf{u} \mapsto \mathbf{y}(\mathbf{u}), \mathbf{z}(\mathbf{u})$ are also differentiable (at least in such a vicinity of $\overline{\mathbf{u}}$). On the other hand, $G(\mathbf{u}) = f(\mathbf{x}(\mathbf{u}))$. By Lemma 3.1,

$$\nabla G(\overline{\mathbf{u}}) \cdot \overline{\mathbf{u}} = 0, \quad \nabla G(\overline{\mathbf{u}}) \le 0, \overline{\mathbf{u}} \le 0. \tag{3.8}$$

But

$$\nabla G(\overline{\mathbf{u}}) = \nabla f(\mathbf{x}(\overline{\mathbf{u}}))\nabla \mathbf{x}(\overline{\mathbf{u}}). \tag{3.9}$$

Multiplying (3.6) by the differential $\nabla \mathbf{x}(\overline{\mathbf{u}})$, and differentiating (3.7) with respect to $\mathbf{u}$, we arrive at

$$\nabla f(\mathbf{x}(\overline{\mathbf{u}}))\nabla \mathbf{x}(\overline{\mathbf{u}}) + \mathbf{y}(\overline{\mathbf{u}})\nabla \mathbf{h}(\mathbf{x}(\overline{\mathbf{u}}))\nabla \mathbf{x}(\overline{\mathbf{u}}) + \mathbf{z}(\overline{\mathbf{u}})\nabla \mathbf{g}(\mathbf{x}(\overline{\mathbf{u}}))\nabla \mathbf{x}(\overline{\mathbf{u}}) = \mathbf{0},$$

$$\nabla \mathbf{h}(\mathbf{x}(\overline{\mathbf{u}}))\nabla \mathbf{x}(\overline{\mathbf{u}}) = \mathbf{0}, \quad \nabla \mathbf{g}(\mathbf{x}(\overline{\mathbf{u}}))\nabla \mathbf{x}(\overline{\mathbf{u}}) = 1.$$

We deduce that

$$\nabla f(\mathbf{x}(\overline{\mathbf{u}}))\nabla \mathbf{x}(\overline{\mathbf{u}}) + \mathbf{z}(\overline{\mathbf{u}}) = \mathbf{0},$$

and then, we identify through (3.9)

$$\nabla G(\overline{\mathbf{u}}) = -\mathbf{z}(\overline{\mathbf{u}}). \tag{3.10}$$

If we put

$$\overline{\mathbf{z}} = \mathbf{z}(\overline{\mathbf{u}}), \quad \overline{\mathbf{y}} = \mathbf{y}(\overline{\mathbf{u}}),$$

and recall that $\overline{\mathbf{u}} = \mathbf{g}(\overline{\mathbf{x}})$, the identification (3.10), together with (3.6)–(3.8), proves the theorem.

Let us discuss a couple of simple examples to see the kind of calculations we are talking about. Our readers will soon realize that even for rather moderate-sized problems computations by hand are impossible.

Example 3.4 Consider the following situation

$$\text{Minimize in } (x_1, x_2): \quad x_1^2 + 2x_2^2 - x_1(30 - x_1) - x_2(35 - x_2)$$

under

$$x_1^2 + 2x_2^2 \le 250, \quad x_1 + x_2 \le 16, \quad x_1, x_2 \ge 0.$$

Fig. 3.3 The feasible set for
Example 3.4

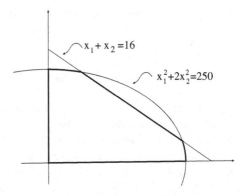

Because we have two independent variables, it is not hard to visualize the feasible set in $\mathbb{R}^2$, and conclude that it is a bounded set. A sketch of the feasible set has been drawn in Fig. 3.3. By the Weierstrass criterium the problem does admit a global minimum. On the other hand, the cost function can be simplified to $2x_1^2 + 3x_2^2 - 30x_1 - 35x_2$.

Because we are supposed to respect four inequalities, we need to introduce four multipliers z_i, $i = 1, 2, 3, 4$. Optimality conditions are obtained by differentiating the lagrangian

$$2x_1^2 + 3x_2^2 - 30x_1 - 35x_2 + z_1(x_1^2 + 2x_2^2 - 250) + z_2(x_1 + x_2 - 16) - z_3x_1 - z_4x_2$$

with respect to x_i, and equaling them to zero; and by demanding that the four products

$$z_1(x_1^2 + 2x_2^2 - 250) = z_2(x_1 + x_2 - 16) = z_3x_1 = z_4x_2 = 0 \qquad (3.11)$$

vanish. Among all possible solutions we are only interested in those that comply with

$$z_i \geq 0, \quad x_1^2 + 2x_2^2 \leq 250, \quad x_1 + x_2 \leq 16, \quad x_1, x_2 \geq 0. \qquad (3.12)$$

The equations for the partial derivatives read

$$4x_1 - 30 + 2x_1z_1 + z_2 - z_3 = 0, \quad 6x_2 - 35 + 2x_2z_1 + z_2 - z_4 = 0. \qquad (3.13)$$

The products in (3.11) enable us to go through the following enumeration of sixteen possibilities:

1. $z_1 = z_2 = z_3 = z_4 = 0$. In this case, (3.13) imply that $x_1 = 15/2, x_2 = 35/6$; in order to consider this point as a valid candidate we need to ensure that it respects (3.12). It does indeed, and we have one candidate to compete for the minimum.
2. $z_1 = z_2 = z_3 = x_2 = 0$. Looking at (3.13), we conclude that $x_1 = 15/2$, but the second equation yields $z_4 = -35$ which cannot correspond to a minimum.
3. $z_1 = z_2 = x_1 = z_4 = 0$. The first equation in (3.13) yields immediately $z_3 = -30$, and this cannot be so.

4. $z_1 = z_2 = x_1 = x_2 = 0$. Both $z_3 = -30$, and $z_4 = -35$ are negative.
5. $z_1 = x_1 + x_2 - 16 = z_3 = z_4 = 0$. By eliminating z_2 from the two equations (3.13), we arrive at

$$4x_1 - 6x_2 = -5, \quad x_1 + x_2 = 16,$$

and, hence, $x_1 = 9.1$, $x_2 = 6.9$. But then $z_2 = 30 - 4x_1 < 0$, and this case do not furnish any candidate either.
6. $z_1 = x_1 + x_2 - 16 = z_3 = x_2 = 0$. $x_1 = 16$, and, as before, $z_2 = 30 - 4x_1 < 0$.
7. $z_1 = x_1 + x_2 - 16 = x_1 = z_4 = 0$. $x_2 = 16$, and $z_2 = 35 - 6x_2 < 0$.
8. $z_1 = x_1 + x_2 - 16 = x_1 = x_2 = 0$. Incompatible.
9. $x_1^2 + 2x_2^2 - 250 = z_2 = z_3 = z_4 = 0$. This case is not so easily discarded as the explicit computations, though looking simple, are not so. However there is no feasible solution either.
10. $x_1^2 + 2x_2^2 - 250 = z_2 = z_3 = x_2 = 0$. The only feasible point is $x_1 = \sqrt{250}$, $x_2 = 0$, but this time again $z_4 = -35$, and no candidate has to be considered.
11. $x_1^2 + 2x_2^2 - 250 = z_2 = x_1 = z_4 = 0$. Again this time $z_3 = -30$.
12. $x_1^2 + 2x_2^2 - 250 = z_2 = x_1 = x_2 = 0$. Incompatible.
13. $x_1^2 + 2x_2^2 - 250 = x_1 + x_2 - 16 = z_3 = z_4 = 0$. The equation for x_2 is the quadratic one $3x_2^2 - 32x_2 + 6 = 0$ with two positive roots, however the values of z_1 and/or z_2 are negative, and no point is obtained in this case.
14. $x_1^2 + 2x_2^2 - 250 = x_1 + x_2 - 16 = z_3 = x_2 = 0$. Incompatible.
15. $x_1^2 + 2x_2^2 - 250 = x_1 + x_2 - 16 = x_1 = z_4 = 0$. Incompatible.
16. $x_1^2 + 2x_2^2 - 250 = x_1 + x_2 - 16 = x_1 = x_2 = 0$. Incompatible.

After the full treatment of the sixteen cases, the only candidate is the one in the first case $(15/2, 35/6)$ which is indeed the minimum sought.

This simple example which only involves the simplest non-linear functions (quadratics) already informs us that the matter of finding the optimal solutions of non-linear mathematical programs can be quite delicate and cumbersome. In addition to the full enumeration of all possibilities, which is somewhat tedious, to ensure that we do not miss the true solution, we may find some of the cases hard to analyze even for the simplest situations. Most of them might be very easily dealt with though. We will learn in the next section how important, even from a practical viewpoint, is to have in advance important information about the problem.

Example 3.5 This time we would like to find the two extreme, global minimum and maximum, for the cost function

$$f(x_1, x_2, x_3) = x_1^2 + x_2^2 + x_3^2 + x_1 + x_2 + x_3$$

over the feasible set determined by the constraints

$$x_1^2 + x_2^2 + x_3^2 = 4, \quad x_3 \leq 1.$$

Fig. 3.4 The feasible set for
Example 3.5

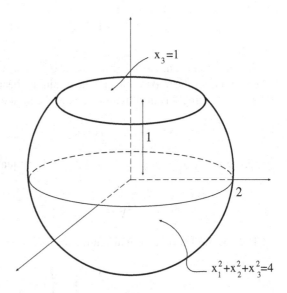

$x_3=1$

1

2

$x_1^2+x_2^2+x_3^2=4$

It is easy to figure out what the feasible set is: the part of the sphere centered at the origin with radius 2 under the horizontal plane $x_3 = 1$, Fig. 3.4. It is definitely a bounded set, and so, by the Weierstrass criterium, there are at least two points where the maximum, and the minimum are taken on. Because the maximum is always the minimum with an extra minus sign, this translates into the fact that candidates for the maximum are those solutions of the Karush–Kuhn–Tucker (KKT henceforth) conditions for which $z \leq 0$. Note that points where $z = 0$ are to be considered in both lists: the one for the maximum, and the one for the minimum. In general, critical points for which there is no coherence in the signs of the multipliers $\mathbf{z}$ (some components are positive and some other negative) cannot be eligible candidates either for the maximum or the minimum.

This time optimality conditions read

$$2x_1 + 1 + 2yx_1 = 0, \quad 2x_2 + 1 + 2yx_2 = 0, \quad 2x_3 + 1 + 2yx_3 + z = 0,$$
$$z(x_3 - 1) = 0, \quad x_1^2 + x_2^2 + x_3^2 = 4.$$

We will have to put solutions in two lists: those with $z \geq 0$ for the minimum, and those with $z \leq 0$ for the maximum. We only have one equation in the form of a product (one single inequality constraint), and so the discussion this time proceeds in two possible cases.

1. $z = 0$. This time the system reduces to

$$2(1 + y)x_1 = 2(1 + y)x_2 = 2(1 + y)x_3 = -1.$$

Since $1 + y$ cannot vanish, we conclude that $x_1 = x_2 = x_3$, and the equality constraint lead to the two points

$$\frac{2}{\sqrt{3}}(1, 1, 1), \quad -\frac{2}{\sqrt{3}}(1, 1, 1).$$

The first point is infeasible (because $2/\sqrt{3} > 1$). The other one has to be considered both for the maximum, and the minimum because the multiplier z vanishes.

2. $x_3 = 1$. Taking this information back to the system, we find that

$$2(1 + y)x_1 = 2(1 + y)x_2 = -1, \quad z = -3 - 2y.$$

Hence $x_1 = x_2, x_3 = 1, z = -1 + 1/x_1$. The equality constraints yield

$$x_1 = x_2 = \pm\sqrt{\frac{3}{2}}, \quad z = -1 \pm \sqrt{\frac{2}{3}} < 0 \text{ (in both cases).}$$

Then we find two additional candidates for the maximum in the two points

$$\left(\sqrt{\frac{3}{2}}, \sqrt{\frac{3}{2}}, 1\right), \quad \left(-\sqrt{\frac{3}{2}}, -\sqrt{\frac{3}{2}}, 1\right).$$

We therefore have the points:

1. minimum:

$$-\frac{2}{\sqrt{3}}(1, 1, 1);$$

2. maximum:

$$-\frac{2}{\sqrt{3}}(1, 1, 1), \left(\sqrt{\frac{3}{2}}, \sqrt{\frac{3}{2}}, 1\right), \left(-\sqrt{\frac{3}{2}}, -\sqrt{\frac{3}{2}}, 1\right).$$

The minimum is definitely attained at

$$-\frac{2}{\sqrt{3}}(1, 1, 1),$$

while the maximum takes place at

$$\left(\sqrt{\frac{3}{2}}, \sqrt{\frac{3}{2}}, 1\right),$$

once the cost function f is evaluated at the three candidates for the maximum. A final remark is that because the equality constraint $x_1^2 + x_2^2 + x_3^2 = 4$ is to be enforced, one can take this information to the cost function and conclude that our problem is equivalent to finding the extreme points for the new cost function $\tilde{f}(x_1, x_2, x_3) =$

$x_1 + x_2 + x_3$ under the same constraints. In this way, some computations may be simplified by considering this new cost function $\tilde{f}$.

Example 3.6 We finally examine optimality conditions for a typical linear programming problem, as we had promised in the linear-programming chapter,

$$\text{Maximize in } \mathbf{x}: \quad \mathbf{c} \cdot \mathbf{x} \quad \text{under} \quad \mathbf{Ax} \leq \mathbf{b}, \mathbf{x} \geq \mathbf{0}.$$

Upon reflecting a little bit on the situation, optimality conditions can be written with the help of two vectors of multipliers $\mathbf{y} \in \mathbb{R}^m$, $\mathbf{z} \in \mathbb{R}^N$, namely,

$$-\mathbf{c} + \mathbf{yA} - \mathbf{z} = \mathbf{0}, \quad \mathbf{y}(\mathbf{Ax} - \mathbf{b}) = 0, \quad \mathbf{z} \cdot \mathbf{x} = 0,$$

in addition to

$$\mathbf{y} \geq \mathbf{0}, \quad \mathbf{z} \geq \mathbf{0}, \quad \mathbf{Ax} \leq \mathbf{b}, \quad \mathbf{x} \geq \mathbf{0}.$$

The minus in front of vector $\mathbf{c}$ accounts for the fact that we are seeking a maximum instead of a minimum. These are exactly the conditions we used for duality in linear programming in the last chapter.

3.4 Convexity

Readers may feel disappointed at seeing the great effort that requires solving non-linear programming problems. There is no way to anticipate how many solutions are expected, where they are expected (other than at the solution set of optimality conditions), or if they will be local or global. In fact, as pointed out earlier in the text, the non-linear world is so complex that there is apparently no way to set up some order or rule that may help us in knowing where we are, and what to expect. In the middle of the chaos of the non-linear world, convexity stands to convey some hope.

To motivate the concept of convexity, and its importance, let us go back to the simplest, non-trivial programming problem one can think of

$$\text{Minimize in } x \in \mathbb{R}: \quad f(x), \tag{3.14}$$

where as usual, we assume f to be as smooth as we may need it to be. The first step is to ensure that such a problem admits optimal solutions, i.e. there is some real number $\overline{x}$ with the property $f(\overline{x}) \leq f(x)$ for every $x \in \mathbb{R}$. Because the real line $\mathbb{R}$ is not a bounded set, we will be using the Weierstrass criterium, Proposition 3.1, for unbounded sets so that we have

$$\lim_{x \to \pm\infty} f(x) = +\infty. \tag{3.15}$$

This "coercivity" hypothesis convinces us that, indeed, there has to be at least one point $\bar{x}$ of minimum for f. If this condition does not hold, the point of minimum could travel all the way to infinity, and there might not be a minimizer, though some times there might one.

Once we are sure that there are minimizers, we set to ourselves the task of finding them, or at least one of them. From elementary Calculus, we know that such minimizers will have to hide among the critical points of f, that is to say, among the roots of its derivative $f'(x) = 0$. We are also well aware that there may be many other critical points, which are solutions of $f'(x) = 0$, but are not the points we are after: local minima, local maxima, saddle points, etc. Once we would find all those, there is the job to discern which one(s) are the global minimizer(s) for our program. This task may not be so easy some times, even if we assume that we can write a full list of all those critical points. The question is: can there possibly be some kind of structural property on f that could save us the worry to go through this final discernment step? Said differently, is there a property on f that could be a sure indication or certificate that every critical point is a minimizer? Or even better, that there cannot be more than one critical point?

The answer, at least for our simple mathematical program, is not that difficult. Suppose we have found a certain critical point x_1, so that $f'(x_1) = 0$. We would like to discard the possibility of having a different x_2, say $x_2 > x_1$, with $f'(x_2) = 0$. An appealing intuitive idea is to demand that once f' has passed through zero at x_1, then it cannot go back to zero again. This can be accomplished by asking for the monotonicity of f'. But we know from Calculus that a smooth function is monotone if its derivative, this time f'', never vanishes !! This is equivalent to saying that it has constant sign. The sign that is compatible with the coercivity above is positive.

Proposition 3.2 *Suppose* $f : \mathbb{R} \to \mathbb{R}$ *is smooth.*

1. *If f is coercive in the sense (3.15), then (3.14) admits at least one optimal (global) solution.*
2. *If, in addition, $f''(x) \geq 0$ for all $x \in \mathbb{R}$, then every critical point x, $f'(x) = 0$, is a (global) solution of (3.14).*
3. *If, in addition, $f'(x) > 0$ for every $x \in \mathbb{R}$, then there is just one unique critical point for f which is the unique (global) solution of (3.14).*

This interesting discussion leads to the concept of convexity.

Definition 3.2 A smooth function $f : \mathbb{R} \to \mathbb{R}$ is convex if $f'' \geq 0$. If, in addition, $f'' > 0$, then f is said to be strictly convex.

We need to do more work in two important directions: what convexity is for functions $f : \mathbb{R}^N \to \mathbb{R}$ of several variables, and to state the relevance of convexity for a general non-linear mathematical program.

We do not pretend here to write a small formal treatise on convexity. Our modest aim is to state the fundamental properties of convex functions as they relate to non-linear mathematical programming problems.

Definition 3.3 A function $f : \mathbb{R}^N \to \mathbb{R}$ is convex if

$$f(t\mathbf{x}_1 + (1 - t)\mathbf{x}_0) \leq t f(\mathbf{x}_1) + (1 - t) f(\mathbf{x}_0)$$

for all $t \in [0, 1]$, and all pairs of vectors $\mathbf{x}_1, \mathbf{x}_0 \in \mathbb{R}^N$.

The expression $t\mathbf{x}_1 + (1 - t)\mathbf{x}_0$ for $t \in [0, 1]$ is called a convex combination of the two vectors. When t runs through $[0, 1]$, the convex combination covers the full interval $[\mathbf{x}_0, \mathbf{x}_1]$. The convexity of f means that its value at any given convex combination is always no greater than the same convex combination of the values of f at the end-points. Alternatively, one can say that the graph of f is always under any of its secants.

When we demand more smoothness on f, the convexity can be, equivalently, expressed by making use of derivatives.

Proposition 3.3 *1. If $f : \mathbb{R}^N \to \mathbb{R}$ has first partial derivatives which are continuous, then f is convex if and only if*

$$f(\mathbf{y}) \geq f(\mathbf{x}) + \nabla f(\mathbf{x}) \cdot (\mathbf{y} - \mathbf{x})$$

for all pairs $\mathbf{x}, \mathbf{y} \in \mathbb{R}^N$.
2. If $f : \mathbb{R}^N \to \mathbb{R}$ has up to second partial derivatives which are continuous, then f is convex if and only if the (symmetric) hessian matrix $\nabla^2 f(\mathbf{x})$ has non-negative eigenvalues at every $\mathbf{x} \in \mathbb{R}^N$. Equivalently, if $\nabla^2 f(\mathbf{x})$ is positive semidefinite for every such $\mathbf{x}$; or, from the well-known criterium involving the principal subdeterminants, if the principal subdeterminants of the hessian $\nabla^2 f(\mathbf{x})$ are non-negative.

From a practical point of view, and assuming that most of the time functions occurring in mathematical programs are smooth, the best criterium for convexity is the one involving the hessian. This does not mean that checking convexity cannot be really tough some times.

There is also the important concept of strict convexity, which is also very convenient when we are interested in knowing when optimal solutions are unique.

Definition 3.4 A function $f : \mathbb{R}^N \to \mathbb{R}$ is strictly convex if it is convex, and

$$f(t\mathbf{x}_1 + (1-t)\mathbf{x}_0) = tf(\mathbf{x}_1) + (1-t)f(\mathbf{x}_0)$$

can only happen if $t \in \{0, 1\}$, or else when $\mathbf{x}_0 = \mathbf{x}_1$.

As before, there is also a criterium for strict convexity based on derivatives.

Proposition 3.4 *1. If $f : \mathbb{R}^N \to \mathbb{R}$ has first partial derivatives which are continuous, then f is strictly convex if*

$$f(\mathbf{y}) \geq f(\mathbf{x}) + \nabla f(\mathbf{x}) \cdot (\mathbf{y} - \mathbf{x})$$

for all pairs $\mathbf{x}, \mathbf{y} \in \mathbb{R}^N$, and equality can only happen for $\mathbf{y} = \mathbf{x}$.
2. If $f : \mathbb{R}^N \to \mathbb{R}$ has up to second partial derivatives which are continuous, then f is strictly convex if the hessian matrix $\nabla^2 f(\mathbf{x})$ has strictly positive eigenvalues at every $\mathbf{x} \in \mathbb{R}^N$. Equivalently, if $\nabla^2 f(\mathbf{x})$ is positive definite for every such $\mathbf{x}$; or, if the leading subdeterminants of the hessian $\nabla^2 f(\mathbf{x})$ are strictly positive.

The prototype of a convex function which is not strictly convex is a linear function. A linear function f is such that both f and $-f$ are convex. A function is concave if $-f$ is convex, so that linear functions are, at the same time, convex and concave.

Understanding and appreciating the importance of convexity requires quite an amount of work. Anyhow, its relevance for non-linear programming problems can be stated as follows. We go back to a typical problem like (3.1)

$$\text{Minimize in } \mathbf{x} \in \mathbb{R}^N : \quad f(\mathbf{x}) \quad \text{subject to} \quad \mathbf{h}(\mathbf{x}) = \mathbf{0}, \mathbf{g}(\mathbf{x}) \leq \mathbf{0}, \qquad (3.16)$$

where we suppose that all functions involved are smooth, and that the problem does admit global minimizers.

Theorem 3.2 (Sufficiency of KKT)

1. Let us assume that f, and all components of **g** are convex functions, and all components of **h** are linear. Then every solution of the KKT optimality conditions is an optimal solution of problem (3.17).
2. If, in addition, f is strictly convex, then there cannot be more than one solution of the KKT optimality conditions, and this solution is the unique minimizer of the problem.

Proof Suppose $\overline{\mathbf{x}}$, together with $(\overline{\mathbf{y}}, \overline{\mathbf{z}})$, is indeed a solution of the KKT optimality conditions so that

$$\nabla f(\overline{\mathbf{x}}) + \overline{\mathbf{y}}\nabla\mathbf{h}(\overline{\mathbf{x}}) + \overline{\mathbf{z}}\nabla\mathbf{g}(\overline{\mathbf{x}}) = \mathbf{0},$$
$$\overline{\mathbf{z}} \cdot \mathbf{g}(\overline{\mathbf{x}}) = 0, \quad \mathbf{h}(\overline{\mathbf{x}}) = \mathbf{0},$$
$$\overline{\mathbf{z}} \geq \mathbf{0}, \quad \mathbf{g}(\overline{\mathbf{x}}) = \mathbf{0}.$$

Let us argue, first, that $\overline{\mathbf{x}}$ is a local point of minimum. Suppose not, and let $\tilde{\mathbf{x}}$ be a feasible point for (3.17), close to $\overline{\mathbf{x}}$, such that $f(\tilde{\mathbf{x}}) < f(\overline{\mathbf{x}})$. The so-called lagrangian L of the problem

$$L(\mathbf{x}) = f(\mathbf{x}) + \overline{\mathbf{y}}\mathbf{h}(\mathbf{x}) + \overline{\mathbf{z}}\mathbf{g}(\mathbf{x})$$

cannot have a point of local minimum at $\overline{\mathbf{x}}$, because for every $\tilde{\mathbf{x}}$ as just indicated, we would have

$$L(\tilde{\mathbf{x}}) = f(\tilde{\mathbf{x}}) + \overline{\mathbf{y}} \cdot \mathbf{h}(\tilde{\mathbf{x}}) + \overline{\mathbf{z}} \cdot \mathbf{g}(\tilde{\mathbf{x}}) \leq f(\tilde{\mathbf{x}}) < f(\overline{\mathbf{x}}) = L(\overline{\mathbf{x}}),$$

precisely because

$$\mathbf{h}(\tilde{\mathbf{x}}) = \mathbf{h}(\overline{\mathbf{x}}) = \mathbf{0}, \quad \mathbf{g}(\tilde{\mathbf{x}}) \leq \mathbf{0}, \overline{\mathbf{z}} \geq \mathbf{0}, \quad \overline{\mathbf{z}} \cdot \mathbf{g}(\overline{\mathbf{x}}) = 0.$$

Yet

$$\nabla L(\overline{\mathbf{x}}) = \nabla f(\overline{\mathbf{x}}) + \overline{\mathbf{y}}\nabla\mathbf{h}(\overline{\mathbf{x}}) + \overline{\mathbf{z}}\nabla\mathbf{g}(\overline{\mathbf{x}}) = \mathbf{0},$$

which is a contradiction for a convex function like $L(\mathbf{x})$ because convex functions cannot have critical points which are not local minima. Note how $L(\mathbf{x})$ is in fact a convex function under the hypotheses assumed on f, **h**, and **g**.

It is now easy to reinforced the fact that $\overline{\mathbf{x}}$ is, not only a local point of minimum, but also global. This is again a consequence of convexity. If now $\tilde{\mathbf{x}}$ is a feasible point, not necessarily in a vicinity of $\overline{\mathbf{x}}$, where the global minimum is attained, the function $t \in [0, 1] \mapsto f(t\overline{\mathbf{x}} + (1 - t)\tilde{\mathbf{x}})$ of a single variable t is convex, and has minimum points at both $t = 0, 1$. The only possibility is that it is constant, and $\overline{\mathbf{x}}$ is also a point of global minimum, i.e. an optimal solution of (3.17).

It is elementary to strengthen this argument to conclude the uniqueness of global solution based on the strict convexity of f.

The significance of this result cannot be missed. What we are asserting is that, under convexity (and linearity) conditions for the functions involved, which is something that can be checked beforehand, KKT optimality conditions are indeed the clue to finding optimal solutions, and there is no need of further discernment about whether they are or they are not the solutions we are after, because they have to be. Moreover, if the strict convexity for the cost function f is also a fact, then once we have found one good candidate, a good solution of optimality conditions, there is no need to look for more for they will be no more.

It is very easy to check that all of these convexity and linearity hypotheses do hold for Example 3.4, including the strict convexity of the cost function f. Therefore, according to Theorem 3.2, once the first point $(15/2, 35/6)$ is found in the first case of the sixteen that were examined, there is no need to go through the other fifteen cases because convexity guarantees that another candidate will not be found, even if some of the potential candidates cannot be written or found explicitly. Note, that, because a maximum problem can be transformed into a minimum problem by a change of sign of the objective function, a change of sign in a convex function gives rise to a concave function, which is the right concept for maximization problems. No further comments are necessary.

For Example 3.5, Theorem 3.2 cannot be used because the only equality constraint is setup with a non-linear function. In this case, there is no substitute for finding, as we did, a full list of candidates for the maximum, for the minimum, and discern which ones are a global minimum, and a global maximum by evaluating the cost function.

Finally, for linear programs Theorem 3.2 always holds. Indeed, this is the basis of duality in linear programming as we argued in Chap. 2.

3.5   Duality

Duality in non-linear programming is much more involved than in its linear counterpart. Yet, it is fundamental to appreciate the deep structure of non-linear programming. We just pretend in this short section to introduce the dual function, and relate it to the KKT-optimality conditions. A more profound analysis is deferred to a more advanced course.

Let

Minimize in $\mathbf{x} \in \mathbb{R}^N$: $f(\mathbf{x})$ subject to $\mathbf{h}(\mathbf{x}) = \mathbf{0}, \mathbf{g}(\mathbf{x}) \leq \mathbf{0}$,   (3.17)

be our primal problem. The associated lagrangian was already introduced as

$$L(\mathbf{x}; \mathbf{y}, \mathbf{z}) = f(\mathbf{x}) + \mathbf{y} \cdot \mathbf{h}(\mathbf{x}) + \mathbf{z} \cdot \mathbf{g}(\mathbf{x})$$

where we now want to stress its dependence on the three sets of variables $(\mathbf{x}; \mathbf{y}, \mathbf{z})$. Optimality conditions can then be expressed as those solutions of the non-linear system

$$\nabla_{\mathbf{x}} L(\mathbf{x}; \mathbf{y}, \mathbf{z}) = \mathbf{0}, \quad \mathbf{z} \cdot \mathbf{g}(\mathbf{x}) = 0, \quad \mathbf{h}(\mathbf{x}) = \mathbf{0}, \tag{3.18}$$

which comply with the sign constraints

$$\mathbf{g}(\mathbf{x}) \leq \mathbf{0}, \quad \mathbf{z} \geq \mathbf{0}.$$

Suppose we could solve for $\mathbf{x}$ in terms of $(\mathbf{y}, \mathbf{z})$ in the system

$$\nabla_{\mathbf{x}} L(\mathbf{x}; \mathbf{y}, \mathbf{z}) = \mathbf{0},$$

in a unique way, and write $\mathbf{x}(\mathbf{y}, \mathbf{z})$ to express that dependence of the point of minimum $\mathbf{x}$ upon multipliers $(\mathbf{y}, \mathbf{z})$. We therefore have

$$\nabla_{\mathbf{x}} L(\mathbf{x}(\mathbf{y}, \mathbf{z}); \mathbf{y}, \mathbf{z}) = \mathbf{0}. \tag{3.19}$$

Definition 3.5 The function

$$\phi(\mathbf{y}, \mathbf{z}) = L(\mathbf{x}(\mathbf{y}, \mathbf{z}); \mathbf{y}, \mathbf{z})$$
$$= f(\mathbf{x}(\mathbf{y}, \mathbf{z})) + \mathbf{y} \cdot \mathbf{h}(\mathbf{x}(\mathbf{y}, \mathbf{z})) + \mathbf{z} \cdot \mathbf{g}(\mathbf{x}(\mathbf{y}, \mathbf{z}))$$

is the dual function of problem (3.17), and the dual problem itself is

Maximize in $(\mathbf{y}, \mathbf{z}) \in \mathbb{R}^m \times \mathbb{R}^n$: $\phi(\mathbf{y}, \mathbf{z})$

under $\mathbf{z} \geq \mathbf{0}$.

Note that, under the same convexity/linearity hypotheses as in Theorem 3.2, the dual function is nothing but the minimum of the lagrangian with respect to $\mathbf{x}$ for each fixed $(\mathbf{y}, \mathbf{z})$

$$\phi(\mathbf{y}, \mathbf{z}) = \min_{\mathbf{x}} L(\mathbf{x}; \mathbf{y}, \mathbf{z}),$$

and so (3.19) holds.

Two relevant remarks to emphasize how this problem relates to the original one.

1. Assume the triplet $(\mathbf{x}, \mathbf{y}, \mathbf{z})$ is such that

$$\mathbf{z} \geq \mathbf{0}, \quad \mathbf{h}(\mathbf{x}) = \mathbf{z} \cdot \mathbf{g}(\mathbf{x}) = \mathbf{0}.$$

Then

$$f(\mathbf{x}) = f(\mathbf{x}) + \mathbf{y} \cdot \mathbf{h}(\mathbf{x}) + \mathbf{z} \cdot \mathbf{g}(\mathbf{x}) \geq \phi(\mathbf{y}, \mathbf{z})$$

because the dual function is defined precisely by minimizing the lagrangian with respect to $\mathbf{x}$ for each fixed pair $(\mathbf{y}, \mathbf{z})$. The arbitrariness of the triplet under those conditions lead immediately to the inequality

$$\max \phi(\mathbf{y}, \mathbf{z}) \leq \min f(\mathbf{x}).$$

Equality takes place essentially when KKT optimality conditions are sufficient under convexity and linearity conditions (Theorem 3.2). Strict inequality is identified, in the jargon of mathematical programming, as the "duality gap".

2. KKT optimality conditions for both problems, the primal and its dual, take us to the same underlying set of conditions. Indeed, if we put $\mathbf{x} \equiv \mathbf{x}(\mathbf{y}, \mathbf{z})$ for a solution of (3.18), then some careful manipulations, bearing in mind the chain rule for differentiation of compositions of maps assuming these are differentiable, yield that

$$\nabla_{\mathbf{y}}\phi = \nabla f(\mathbf{x})\nabla_{\mathbf{y}}\mathbf{x} + \mathbf{h}(\mathbf{x}) + \mathbf{y}\nabla\mathbf{h}(\mathbf{x})\nabla_{\mathbf{y}}\mathbf{x} + \mathbf{z}\nabla\mathbf{g}(\mathbf{x})\nabla_{\mathbf{y}}\mathbf{x} = \mathbf{h}(\mathbf{x}).$$

Similarly

$$\nabla_{\mathbf{z}}\phi = \nabla f(\mathbf{x})\nabla_{\mathbf{z}}\mathbf{x} + \mathbf{y}\nabla\mathbf{h}(\mathbf{x})\nabla_{\mathbf{z}}\mathbf{x} + \mathbf{z}\nabla\mathbf{g}(\mathbf{x})\nabla_{\mathbf{z}}\mathbf{x} + \mathbf{g}(\mathbf{x}) = \mathbf{g}(\mathbf{x}).$$

These are exactly optimality conditions for the dual problem. Indeed, KKT optimality conditions for the dual problem

$$\text{Maximize in } (\mathbf{y}, \mathbf{z}) \in \mathbb{R}^m \times \mathbb{R}^n : \quad \phi(\mathbf{y}, \mathbf{z})$$

under $\mathbf{z} \geq \mathbf{0}$, lead immediately to

$$\nabla_{\mathbf{y}}\phi(\mathbf{y}, \mathbf{z}) = \mathbf{0}, \quad \nabla_{\mathbf{z}}\phi(\mathbf{y}, \mathbf{z}) - \mathbf{X} = \mathbf{0}, \quad \mathbf{X} \leq \mathbf{0},$$

if $\mathbf{X}$ is the vector of multipliers corresponding to the sign constraints $\mathbf{z} \geq \mathbf{0}$. The previous computations imply then that

$$\mathbf{h}(\mathbf{x}) = \mathbf{0}, \quad \mathbf{g}(\mathbf{x}) \leq \mathbf{0}.$$

Finally, we just state, without a formal proof though the ideas have been already discussed, a typical duality result for non-linear programming under convexity/linearity constraints.

Theorem 3.3 *A triplet* $(\mathbf{x}, \mathbf{y}, \mathbf{z})$ *is an optimal solution for a primal and a corresponding dual problems under the same hypotheses as in Theorem 3.2 for the functions involved, if and only if*

$$\nabla_{\mathbf{x}} L(\mathbf{x}; \mathbf{y}, \mathbf{z}) = \mathbf{0}, \quad \mathbf{g}(\mathbf{x}) \cdot \mathbf{z} = 0, \quad \mathbf{h}(\mathbf{x}) = \mathbf{0},$$
$$\mathbf{g}(\mathbf{x}) \leq \mathbf{0}, \quad \mathbf{z} \geq \mathbf{0}.$$

3.6 Final Remarks

Non-linear programming is a true jungle. There is no way to anticipate if a given problem of interest will be easy to treat, how many solutions there will be, where they are supposed to be found, etc. Unless we check a priori those convexity and linearity properties, there is no way to know if a given candidate is the global solution sought, or if there will be a better solution somewhere else within the feasible region. There is no way to be certain of having a global solution, even if it is !! It is a dramatic situation. Convexity (and its variants) is the only hope to set some order into the non-linear world.

On the other hand, convexity is also intimately connected to duality in non-linear programming. We have barely touched on duality. A better and deeper understanding is fundamental both from a more formal analytical viewpoint as well as from a practical perspective. It is reserved for a second round on non-linear programming. Discussions about the various versions of duality theorems, the deeper connection between the primal and the dual, etc., have deliberately left for a more advanced course.

3.7 Exercises

3.7.1 Exercises to Support the Main Concepts

1. Let $\mathbf{A}$ be a symmetric, square matrix. Find the extreme values of the associated quadratic form $\mathbf{x}^*\mathbf{A}\mathbf{x}$ under the constraint $|\mathbf{x}|^2 = 1$.
2. Let $\mathbf{Q}$ be a symmetric, non-singular matrix for which the set of vectors $\mathbf{x}^*\mathbf{Q}\mathbf{x} = 0$ is more than just the zero vector. For another such symmetric matrix $\mathbf{A}$, explore the optimization problem

$$\text{Minimize in } \mathbf{x} \in \mathbb{R}^N : \quad \mathbf{x}^*\mathbf{A}\mathbf{x}$$

   under

$$\mathbf{x}^*\mathbf{Q}\mathbf{x} = 1.$$

   (a) Treat first the particular case where

$$\mathbf{Q} = \begin{pmatrix} 0 & 1 \\ 1 & 0 \end{pmatrix}, \quad \mathbf{A} = \begin{pmatrix} 1 & 0 \\ 0 & 1 \end{pmatrix}.$$

   (b) Argue that the feasible set cannot be bounded.
   (c) Suppose $\mathbf{A}$ is coercive, i.e.

$$\lim_{|\mathbf{x}|\to\infty} \mathbf{x}^*\mathbf{A}\mathbf{x} = +\infty.$$

   Show then that the minimum is attained, and find it.

3. Argue that if $\phi_\gamma(\mathbf{x})$ is a family of convex functions for each index γ in a given index set Γ, then its supremum

$$\phi(\mathbf{x}) = \sup_{\gamma\in\Gamma} \phi_\gamma(\mathbf{x})$$

   is also convex. Apply this method to the family

$$\phi_t(\mathbf{x}) = \frac{1}{2}t^2 \mathbf{a}\cdot\mathbf{x} + t$$

   where $t \in \mathbb{R}$, and $\mathbf{a}$ is a fixed, non-vanishing vector.
4. Check that

$$|\mathbf{x}| = \max_{\mathbf{y}:|\mathbf{y}|=1} \mathbf{x}\cdot\mathbf{y}.$$

   Conclude that the function $|\mathbf{x}|$ is convex.

5. If a function $\psi(\mathbf{x})$ is not convex, one can always defined its convexification by setting
$$\psi^{\sharp}(\mathbf{x}) = \sup\{\phi(\mathbf{x}) : \phi \leq \psi \text{ and } \phi, \text{convex}\}.$$

Show that $\psi^{\sharp}$ is always convex, $\psi^{\sharp}(\mathbf{x}) \leq \psi(\mathbf{x})$ for all $\mathbf{x}$, and that there cannot be a convex function $\eta(\mathbf{x})$ with $\eta(\mathbf{x}) \leq \psi(\mathbf{x})$ for all $\mathbf{x}$, but $\eta(\mathbf{x}_0) > \psi^{\sharp}(\mathbf{x}_0)$ for some $\mathbf{x}_0$. Try to figure out the convexification for a non-convex function $f(x)$ of one variable through a sketch of the graph. If $\psi(x) = x + (x^2 - 1)^2$, find its convexification.

6. Show that if $\psi(\mathbf{x}) : \mathbb{R}^N \rightarrow \mathbb{R}$ is convex, and $f(y) : \mathbb{R} \rightarrow \mathbb{R}$ is convex and increasing in the range of ψ, then the composition $\overline{\psi}(\mathbf{x}) = f(\psi(\mathbf{x}))$ is convex. Apply this result to $\psi(\mathbf{x}) = |\mathbf{x}|$, and $f(y) = y^p$, $p > 1$. Are the resulting functions strictly convex?

7. (a) Find the absolute minimum of the function
$$f(x, y) = \frac{x^n + y^n}{2}$$

subject to $x + y = s$, $x \geq 0$ and $y \geq 0$, where s is a positive constant.

(b) Show the inequality
$$\frac{x^n + y^n}{2} \geq \left(\frac{x + y}{2}\right)^n$$

if $n \geq 1$, $x \geq 0$ and $y \geq 0$.

8. Divide a positive number a into n parts in such a way that the sum of the corresponding squares be minimal. Use this fact to prove the inequality
$$\sum_{i=1}^{n} \frac{x_i^2}{n} \geq \left(\sum_{i=1}^{n} \frac{x_i}{n}\right)^2$$

for an arbitrary vector $\mathbf{x} = (x_1, ..., x_n) \in \mathbb{R}^n$.

9. Prove Hölder's inequality
$$\sum_{k=1}^{n} x_k y_k \leq \left(\sum_{k=1}^{n} x_k^p\right)^{1/p} \left(\sum_{k=1}^{n} y_k^q\right)^{1/q}, \quad \frac{1}{p} + \frac{1}{q} = 1,$$

for $p > 1$, $x_k, y_k > 0$.

10. Use the convexity properties of the log function to show the arithmetic-geometric mean inequality
$$\sum_{i=1}^{n} s_i x_i \geq \prod_{i=1}^{n} (x_i)^{s_i}$$

for $s_i \geq 0$ and $\sum_{i=1}^{n} s_i = 1$.

11. A certain function $x(t)$ is supposed to be a solution of the differential equation

$$x'(t) = x(t)^2(2 - x(t))$$

for all $t \in \mathbb{R}$. Classify the critical values of such a function.

12. Consider the optimization problem

$$\text{Minimize in } (x, y) : \quad x$$

under

$$x^2 \leq y \leq x^3.$$

(a) Argue that the minimum vanishes, and it is attained at $(0, 0)$.

(b) Despite the previous conclusion, check that KKT optimality conditions at $(0, 0)$ do not hold. What is the matter?

3.7.2   Practice Exercises

1. Find the minimum of the function $f(x_1, x_2, x_3) = (1/2)(x_1^2 + x_2^2 + x_3^2)$ over the set of equation $x_1 x_2 + x_1 x_3 + x_2 x_3 = 1$.

2. Find the best parabola through the four points $(-1, 4)$, $(0, 1)$, $(1, 1)$ and $(1, -1)$ by minimizing the least square error.

3. Consider the function

$$f(x, y) = \int_x^y \frac{1}{1 + t^4} \, dt,$$

and find its extreme values over $x^2 + y^2 = 1$.

4. Find the optimal vector for the problem

$$\text{Minimize in } (x_1, x_2) : \quad x_1^2 + x_2^2$$

under

$$x_1 + x_2 \leq 1, \quad x_1 - 2x_2 \leq -1.$$

Same question for the problem

$$\text{Minimize in } (x_1, x_2) : \quad |(x_1, x_2) - (4, 1)|^2$$

under the same constraints.

5. Seek the optimal solution for the quadratic program

$$\text{Minimize in } (x_1, x_2) : \quad x_1^2 + 2x_2^2$$

subjected to

$$x_1^2 + 2x_2 - 1 \leq 0, \quad x_1 - x_2 + 3 = 0.$$

Same question if we change the constraint $x_1 - x_2 + 3 = 0$ to $x_1 - x_2 - 3 = 0$.

6. We are interested in finding the minimum of the linear function $x_1 - x_2 + x_3$ over the set of $\mathbb{R}^3$ determined by the constraints

$$x_1 x_2 + x_1 x_3 + x_2 x_3 \leq 0, \quad x_1^2 + x_2^2 + x_3^2 = 1.$$

(a) Argue that such a minimum value is attained.
(b) Are the three functions involved convex? Explain.
(c) Do sufficiency conditions for KKT hold? Can you guarantee if the optimal solution is unique?
(d) Find the point(s) where the minimum is attained and its value.

7. For the objective function

$$f(x, y, z) = xy + xz + yz,$$

find

(a) the extreme values under $x^2 + y^2 + z^2 = 1$;
(b) the extreme values under $x^2 + y^2 + z^2 = 1$, and $x - y + z = 0$.

8. Find the optimal solution of the program

$$\text{Minimize in } (x, y) : \quad -x + 4y + \frac{1}{4}x^2 + \frac{1}{4}(x - y)^2 \quad \text{under } x, y \geq 0,$$

and the value of the minimum.

9. For the problem

$$\text{Minimize in } (x_1, x_2, x_3) : \quad x_1^2 + x_2^2 + x_3^2$$

under the constraints

$$x_1 + x_2 + x_3 \geq 5, \quad x_2 x_3 \geq 2, \quad x_1 \geq 0, x_2 \geq 0, x_3 \geq 2,$$

write the optimality KKT conditions, and check whether the following three points are eligible candidates for the minimum

$$(3/2, 3/2, 2), \quad (4/3, 2/3, 3), \quad (2, 1, 2).$$

10. For the quadratic cost

$$f(x_1, x_2) = (x_1 - x_2 + 1)^2 + (x_2 - 1)^2,$$

and the three regions

- $R_1 : x_2 \geq 0, 2x_1 + x_2 \leq 2, -2x_1 + x_2 \leq 2$;
- $R_2 : x_2 \leq 2x_1(1 - x_1)$;
- $R_3 : 2x_1 + x_2 \leq 2, -2x_1 + x_2 \geq 2$;

answer the following questions in as precise terms as possible:

(a) argue if f is convex, or even strictly convex;
(b) decide whether the problem of minimizing f over each one of the three regions $R_i, i = 1, 2, 3$, admits optimal solution. When is the optimal solution unique?
(c) find the points where the minimum is attained whenever possible.

11. We would like to find the minimum value of $f(x_1, x_2) = x_1^2 + 2x_2^2 + x_1 x_2$ over the set defined by the inequalities

$$x_1 x_2 \leq 1, \quad x_1^2 + x_2^2 \leq 4.$$

(a) Is the feasible set non-empty?
(b) Why can we be sure that the minimum is attained within the admissible set?
(c) Find the point(s) of minimum, and its value.

12. Consider the problem

$$\text{Minimize in } (x_1, x_2) : \quad \frac{1}{5}x_1 + \frac{1}{6}x_2$$

under

$$\frac{25}{x_1} + \frac{100}{x_2} \leq 12, \quad x_1, x_2 \geq 10.$$

(a) Give three feasible points, and calculate their cost.
(b) Is the feasible region bounded? Argue why or why not.
(c) Show that the problem admits an optimal solution: the infimum is attained.
(d) Check if the functions involved are convex, and conclude if we can be sure that KKT optimality conditions provide the solution we seek.
(e) Find the optimal solution either analytically or graphically.

13. Find the two extreme values (maximum and minimum) of the function

$$f(x, y) = \frac{x^3}{3} - \frac{x^2 y^2}{2} - \frac{x^2}{2} + \frac{y^2}{2}$$

over the set determined by the conditions

$$x^2 + y^2 \leq 4, \quad y \geq x - 1.$$

3.7.3 Case Studies

1. A pharmaceutical company would like to know the best location for a distribution central unit in a given region of influence. If the criterium is to minimize the sum of (the square of) the distances to its three main distribution points in $(0, 0)$, $(30, 50)$, $(100, 30)$, find such an optimal location. If the zone under the condition

$$(x_1 - 50)^2 + (x_2 - 20)^2 \leq 98$$

is occupied by a big park in which the unit cannot be built, what would then the optimal location be?

2. Let us consider the scaffolding system of the previous chapter (Linear Programming) where the maximum stress for cables are

$$A, 200;\ B, 100;\ C, 200;\ D, 100.$$

We would like to maximize the sum of the two weights in equilibrium. The point where the load is applied in the lower bar is not specified, and so it is part of the problem.

   (a) Formulate clearly the corresponding non-linear programming problem.
   (b) Is there a certain change of variables that transform it into a LP situation? If so, do it.
   (c) Find the optimal solution graphically. Write precisely the optimal solution: the maximum load permitted, and the point where the second load must be applied.

3. Find the optimal angle to shoot a bullet with initial velocity v_0 to reach a maximum distance.

4. From a square of cardboard of side L, four equal smaller squares are cut from each corner. Find the side of these smaller squares to maximize the volume of the box built with the resulting cut structure.

5. The Cobb-Douglas utility function is of the form

$$u(x_1, x_2, x_3) = x_1^{\alpha_1} x_2^{\alpha_2} x_3^{\alpha_3}, \quad 0 \leq \alpha_i, \alpha_1 + \alpha_2 + \alpha_3 = 1, x_i \geq 0.$$

If a consumer has resources given by $\bar{x}_i$ for commodity X_i, and their respective prices are p_i, formulate and solve the problem of maximizing satisfaction measured through its utility function.

6. A shipwreck survivor finds himself in a small island with a smooth topography whose altitude is given by the function

$$f(x_1, x_2) = 7 - x_1^2 - 2x_2^2 + x_1 x_2.$$

The island is described by $f(x_1, x_2) \geq 0$, and contains a dangerous volcanic and inaccessible zone limited by

$$2x_1 + x_2 \leq 1/2.$$

Find the best point to place a flag, and so maximize chances to be seen by a ship from afar.

Chapter 4
Numerical Approximation

As most likely readers may have realized, it is not at all reasonable to pretend to find solutions for optimization problems, especially if they reflect a real situation, by hand. It is therefore of paramount importance to address the issue of the practical implementation to find good approximations to mathematical programming problems. There are good solutions to this important problem through MatLab ([23]). In particular, some of the subroutines in the Optimization Toolbox may become an interesting way to solve the issue of the numerical implementation of these optimization problems without paying a lot in terms of time to master the intricacies of the numerical simulation. Subroutines like fmincon, fminunc, linprog can be a reasonable solution. There are also comercial software providing much more efficient and elaborate approximation strategies like GAMS ([17]), Lancelot, CONOPT ([12]), and many other. To enjoy a glimpse of the relevance of the numerical approximation to optimization have a look at [14].

We do not pretend to get lost among all those many possibilities. Our modest intention is to endow readers with the most basic ideas and tools to find reasonably accurate approximated solutions to some simple but not entirely trivial problems. It is also important for us that those ideas can be taken and translated from the finite dimensional case (mathematical programming) to the infinite dimensional situation (variational problems and optimal control) to keep effort to a minimum.

Our strategy aims at being elementary and affordable, but efficient if at all possible. It rests on a procedure to find optimal solutions (minima) of unrestricted problems for a function that depends on parameters that are updated in a iterative procedure until convergence takes place. The basic building block is then a good subroutine to calculate the unrestricted minimum of a function. A good solution for this might be fminunc from MatLab, but one can also write his/her own such code either in MatLab or in other alternatives (Fortran, C, Python, etc.). Because that basic step is so important for us, we will first focus on practical methods of approximation of unrestricted problems, and then move on to explain our main strategy to deal with mathematical programs under constraints.

© Springer International Publishing AG 2017
P. Pedregal, *Optimization and Approximation*, UNITEXT 108,
DOI 10.1007/978-3-319-64843-9_4

4.1 Practical Numerical Algorithms for Unconstrained Minimization

The basic initial question that we want to address is how to find good approximations of the optimal solutions of problems like

$$\text{Minimize in } \mathbf{x} \in \mathbb{R}^N : \quad f(\mathbf{x}),$$

for a given smooth function f (which we assume to be defined in all of $\mathbb{R}^N$). We do know that we need to look among the solutions of the critical point equation (system) $\nabla f(\mathbf{x}) = \mathbf{0}$, but this is precisely what we would like to avoid as this (non-linear) system may be hard to solve, especially when N is large. It may even happen that the gradient is not easily written. Notice that, even in the case that the system is linear, it is unfeasible to solve it by hand as soon as N is four or five.

The standard strategy is easy to understand: if we start out at some vector $\mathbf{x}_0$, called vector of initialization, we would like to move to another one $\mathbf{x}_1$ in such a way that $f(\mathbf{x}_1) < f(\mathbf{x}_0)$, and proceed repeatedly in this fashion until no more decreasing of f is possible. The passage from $\mathbf{x}_0$ to $\mathbf{x}_1$ proceeds in two steps:

1. standing at $\mathbf{x}_0$, choose an appropriate direction $\mathbf{v}$ in which to move to ensure that, in doing so, the value of f will decrease;
2. once $\mathbf{v}$ has being chosen, decide on how far to go from $\mathbf{x}_0$ in that direction $\mathbf{v}$.

As soon as we have a clear and good method to cover these two steps, we proceed iteratively until the procedure no longer updates significantly the underlying vector.

4.1.1 The Direction

The first step involves the notion of descent direction.

Definition 4.1 Let $f : \mathbb{R}^N \to \mathbb{R}$ be a smooth function, and let $\mathbf{x}_0 \in \mathbb{R}^N$.

1. A vector $\mathbf{v}$ is called a descent direction for f at $\mathbf{x}_0$ if the inner product $\langle \nabla f(\mathbf{x}_0), \mathbf{v} \rangle$ is negative.
2. The direction given by $\mathbf{v} = -\nabla f(\mathbf{x}_0)$ is called the steepest-descent direction of f at $\mathbf{x}_0$.

The reason for this concept is quite clear. Let $t \in \mathbb{R}$, and consider the real function $g : \mathbb{R} \to \mathbb{R}$, defined by putting $g(t) = f(\mathbf{x}_0 + t\mathbf{v})$. By the chain rule of Calculus, we have

$$g'(t) = \langle \nabla f(\mathbf{x}_0 + t\mathbf{v}), \mathbf{v} \rangle, \quad g'(0) = \langle \nabla f(\mathbf{x}_0), \mathbf{v} \rangle.$$

Hence if $\mathbf{v}$ is such a descent direction $g'(0) < 0$, and this implies that at least not far from $t = 0$, the values of the function g fall, and so, at least for small positive values of t, if we move from $\mathbf{x}_0$ ($t = 0$) to $\mathbf{x}_1 = \mathbf{x}_0 + t\mathbf{v}$, the values of f will decrease, at least a bit. One then can ask: what is the direction in which the decrease of f at $\mathbf{x}_0$ is as large as possible? This is a main property of the gradient of f at $\mathbf{x}_0$, as it is learnt in Calculus courses. The change of sign indicates that we seek the maximum rate of decay.

We have found a first answer to step 1 above: take $\mathbf{v} = -\nabla f(\mathbf{x}_0)$. If it turns out that we are sitting at a critical point $\mathbf{x}_0$, then $\nabla f(\mathbf{x}_0) = \mathbf{0}$, and we would not move.

Because the gradient of f at $\mathbf{x}_0$ only carries local information about f near $\mathbf{x}_0$, some times choosing as descent direction the opposite of the gradient may be too slow, and not optimal. There are more sophisticated methods to select the direction, the so-called conjugate gradient methods (see [25], for instance), but for the sake of brevity we will not cover those here. For many cases, the gradient choice works pretty well, even though the scheme may be slow at times.

If one is willing to put more effort in computing the descent direction than just approximating or computing the gradient, then a real improvement may be attained. Because our way of approximating a solution of the initial unconstrained problem is to search for a critical point where $\nabla f(\mathbf{x}) = \mathbf{0}$, we can, equivalently, try to tackle the problem

$$\text{Minimize in } \mathbf{x} \in \mathbb{R}^N : \quad \frac{1}{2}|\nabla f(\mathbf{x})|^2.$$

The steepest descent direction for this problem at a given $\mathbf{x}_0$ is $-\nabla f(\mathbf{x}_0)\nabla^2 f(\mathbf{x}_0)$. Since the computation of this direction involves the hessian matrix $\nabla^2 f(\mathbf{x}_0)$, and so it is more expensive, the result is also expected to be better, and so it is. Whenever the hessian matrix is involved in the determination of the descent direction, methods are generally referred to as Newton methods. There are many interesting variants, but we will not discuss them here ([16, 25]).

4.1.2   Step Size

Once, standing at a certain iterate $\mathbf{x}_0$, we have decided on the direction we will move to find the next iterate, we need to determine the size of the step we will take in that direction. Assume $\mathbf{v}$ is the direction we have selected, so that we now face the problem of choosing the value of the real variable $t \geq 0$ for the section $g(t) = f(\mathbf{x}_0 + t\mathbf{v})$. Ideally, we would take t as the place where the derivative of g vanishes:

$$0 = g'(t) = \nabla f(\mathbf{x}_0 + t\mathbf{v}) \cdot \mathbf{v}.$$

However, this is a non-trivial equation to solve, and so we are contented with finding a suitable approximation. After all, we are not interested in finding very accurately this solution, as it is just an intermediate step toward a certain goal. There are many valid ideas to select, in a specific way, the value $\bar{t}$ of t, so that $\mathbf{x}_1 = \mathbf{x}_0 + \bar{t}\mathbf{v}$ is the next iterate in our process.

In choosing a good one, we must bear in mind, since this is just a step in an iterative process which will have to be repeated many times, that the way to select $\bar{t}$ be as cheap as possible, and rely as much as possible on the information we already have. In particular, we know that $g(0) = f(\mathbf{x}_0)$, $g'(0) = \nabla f(\mathbf{x}_0) \cdot \mathbf{x}$. These two pieces of information are not enough to make a reasonable choice as all we could do is to compute the tangent line to g at 0, but this does not suffice to make a good decision for t. We at least need a parabola $h(t) = at^2 + bt + c$ such that

$$c = h(0) = g(0) = f(\mathbf{x}_0), \quad b = g'(0) = \nabla f(\mathbf{x}_0) \cdot \mathbf{v}.$$

But we need some further information to determine the coefficient a. The cheapest additional piece of information is to take some preassigned $t_0 > 0$, and evaluate $g(t_0) = f(\mathbf{x}_0 + t_0\mathbf{v})$. We demand

$$at_0^2 + g'(0)t_0 + g(0) = h(t_0) = g(t_0),$$

and find very easily the corresponding value of a

$$a = \frac{1}{t_0^2}(g(t_0) - g(0) - g'(0)t_0). \tag{4.1}$$

We would like a positive value for a, for otherwise the parabola $h(t)$ would be oriented downwards, and its vertex would correspond to a negative value of t which is not acceptable. If t_0 has been selected so that $g(t_0) - g(0) - g'(0)t_0 < 0$, i.e. the value of g at t_0 is under the tangent line of g at 0, then the value of t_0 is not valid, and it should be changed to a smaller value. Indeed, typically one would ask for a criterium (see Fig. 4.1)

$$g(t_0) > g(0) + cg'(0)t_0, \quad c > 0. \tag{4.2}$$

Fig. 4.1 Selection of step-size

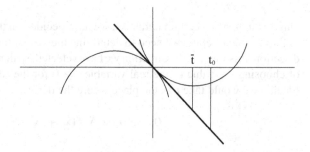

This new parameter c plays the role of demanding a certain rate of farness from the tangent line in the good direction. If the particular value of t_0 that we have selected does not verify (4.2), then we simply update $\mathbf{x}_0$ to $\mathbf{x}_0 + t_0\mathbf{v}$. If it does, then take a as in (4.1), and $\bar{t}$ as the vertex, the point of minimum, of the parabola h, namely

$$\bar{t} = -\frac{b}{2a} = -\frac{1}{2}\frac{t_0^2 \nabla f(\mathbf{x}_0) \cdot \mathbf{v}}{f(\mathbf{x}_0 + t_0\mathbf{v}) - f(\mathbf{x}_0) - \nabla f(\mathbf{x}_0) \cdot \mathbf{v}t_0}.$$

This process of selecting the step size on each iteration is very easily implementable and, at least for standard smooth problems, works reasonably well.

After several successive iterations where descent directions, and suitable step sizes, are chosen in an iterative way until the process stabilizes itself, we will have a fairly good approximation to a candidate for a (local) minimum for f. It f turns out to be convex, or even better strictly convex, that will be an approximation to the true global minimum of f.

We now turn to treat constraints.

4.2 A Perspective for Constrained Problems Based on Unconstrained Minimization

We will start with the simple basic problem

$$\text{Minimize in} \mathbf{x} \in \mathbb{R}^N: \quad f(\mathbf{x}) \quad \text{subject to} \quad g(\mathbf{x}) \leq 0, \tag{4.3}$$

where both f, and g are smooth functions. Let us omit the whole discussion on under what further conditions on both f, and g, our minimization problem may admit optimal solutions. We refer to the previous chapter for some ideas on this. Instead, let us assume that we have gone through all that discussion, and have concluded that indeed we would look for optimal solutions of our problem.

One possibility, indeed quite appealing, would be to directly exploit optimality conditions, namely

$$\nabla f(\mathbf{x}) + \mu \nabla g(\mathbf{x}) = 0, \mu g(\mathbf{x}) = 0, \quad \mu \geq 0, g(\mathbf{x}) \leq 0. \tag{4.4}$$

We do know (under further constraint qualifications) that the vectors we seek should comply with them. However, our viewpoint is even more basic as it does not pretend to enforce, at least on a first attempt, optimality conditions, but directly goes into finding the optimal solutions of the original programming problem. In this way, optimality conditions will be a result of the approximation scheme.

We would like to design an iterative procedure to approximate solutions for (4.3) in a practical, affordable way. We introduce our initial demands in the form:

1. The main iterative step is to be an unconstrained minimization problem, and, as
   such, its corresponding objective function must be defined in all of space (no
   concern whatsoever about feasibility).
2. More specifically, we would like to design a real function G so that the main
   iterative step of our procedure be applied to the augmented cost function
   $L(y, \mathbf{x}) = f(\mathbf{x}) + G(yg(\mathbf{x}))$ for the $\mathbf{x}$-variable. We hope to take advantage of
   the joint dependence upon y and $\mathbf{x}$ inside the argument for G. Notice that letting
   y out of G may not mean a real change as the restriction will again be as in (4.4)
   with a g which would be the composition $G(g(\mathbf{x}))$.
3. The passage from one iterative step to the next is performed through an update
   step for the variable y.
4. Convergence of the scheme ought to lead to a (local) solution of (4.3).

Notice that

$$\nabla_{\mathbf{x}} L(y, \mathbf{x}) = \nabla f(\mathbf{x}) + G'(yg(\mathbf{x}))y\nabla g(\mathbf{x}), \tag{4.5}$$

and so each main iterative step enforces the main equation, the one involving gradients
and derivatives, in (4.4). But we would like to design the function $G(t)$ to ensure
that, as a result of the iterative procedure, the sign conditions in (4.4) are also met.

The following features are to be respected.

1. Variable $\mathbf{x}$ will always be a solution of

$$\nabla f(\mathbf{x}) + G'(yg(\mathbf{x}))y\nabla g(\mathbf{x}) = \mathbf{0}.$$

2. Variable y will always be non-negative (in practice strictly positive but possibly very small). Comparison of (4.5) to (4.4) leads to the identification $\mu =
   G'(yg(\mathbf{x}))y$, and so we would like $G' \geq 0$.
3. The multiplier $\mu = G'(yg(\mathbf{x}))y$ can only vanish if y does. The update rule for
   the variable y will be $y \mapsto G'(yg(\mathbf{x}))y$. If at some step $y = \mu$, the true value of
   the multiplier, then simultaneously $yg(\mathbf{x}) = 0$, and so we would also like to have
   $G'(0) = 1$.
4. If $g(\mathbf{x}) > 0$, then the update rule above for y must yield a higher value for y so as
   to force, in the next iterative step, the feasible inequality $g \leq 0$. Hence, $G'' > 0$,
   or G, convex (for positive values). In addition $G''(t) \to +\infty$ when $t \to +\infty$.
5. If $g(\mathbf{x})$ turns out to be (strictly) negative, then we would like y to become smaller
   so as to let the minimization of f proceed with a lighter interference from the
   inequality constraint. This again leads to G convex (for negative values).
6. The optimality condition $\mu g(\mathbf{x}) = 0$ becomes $G'(yg(\mathbf{x}))yg(\mathbf{x}) = 0$. Thus the
   function $G'(t)t$ can only vanish if $t = 0$. In particular, $G' > 0$.

All of these reasonable conditions impose the requirements

$$G' > 0, G'(0) = 1, G'' > 0, G''(t) \to \infty, \text{if } t \to \infty.$$

Possibly, the most familiar choice if $G(t) = e^t$, and this is the one we will select.

The iterative procedure is then as follows.

1. Initialization. Set $y_0 = 1$, $\mathbf{x}_{-1} = \mathbf{0}$, or any other choice.
2. Iterative step until convergence. Suppose we have y_j, $\mathbf{x}_{j-1}$.
   (a) Solve for the unconstrained optimization problem

   $$\text{Minimize in } \mathbf{z} \in \mathbb{R}^N : \quad f(\mathbf{z}) + e^{y_j g(\mathbf{z})}$$

   starting from the intial guess $\mathbf{x}_{j-1}$. Let $\mathbf{x}_j$ be such (local) minimizer.
   (b) If $y_j g(\mathbf{x}_j)$ vanishes, stop and take $\mathbf{x}_j$ as the solution of the problem.
   (c) If $y_j g(\mathbf{x}_j)$ does not vanish, update y_j by setting $y_{j+1} = e^{y_j g(\mathbf{x}_j)} y_j$.

In order to emphasize the interest of our proposal, let us prove the following simple convergence result.

Theorem 4.1 *Suppose both functions f and g are smooth. If for a sequence of iterates $\{(y_j, \mathbf{x}_j)\}$ produced by our algorithm, it is true that*

$$y_j g(\mathbf{x}_j) \to 0, \quad \mathbf{x}_j \to \tilde{\mathbf{x}},$$

then $\tilde{\mathbf{x}}$ is a local optimal solution of (4.3).

Proof Take $y_0 > 0$ in an arbitrary way, and produce the sequence of iterates $\{(y_j, \mathbf{x}_j)\}$ for our procedure. Suppose, as it has been assumed, that $y_j g(\mathbf{x}_j) \to 0$, and $\mathbf{x}_j \to \tilde{\mathbf{x}}$ for some vector $\tilde{\mathbf{x}}$.

1. We claim that $\tilde{\mathbf{x}}$ is feasible. Suppose not, so that $g(\tilde{\mathbf{x}}) > 0$. Then $g(\mathbf{x}_j) > 0$ for j large enough. In this case

   $$y_{j+1} = e^{y_j g(\mathbf{x}_j)} y_j \geq y_j.$$

   This inequality implies that $\{y_j\}$ is a non-decreasing sequence of positive numbers, at least from some term on. As such it cannot converge to zero, and this is absurd if $y_j g(\mathbf{x}_j) \to 0$ with $g(\tilde{\mathbf{x}}) > 0$.
2. Suppose we could find an admisible vector $\mathbf{x}$ close to $\tilde{\mathbf{x}}$ such that $f(\mathbf{x}) < f(\tilde{\mathbf{x}})$. Then

   $$f(\mathbf{x}) + e^{y_j g(\mathbf{x})} \leq f(\mathbf{x}) + 1 < f(\tilde{\mathbf{x}}) + 1$$

for all j, because $y_j g(\mathbf{x}) \leq 0$. For j sufficiently large, we will then have

$$f(\mathbf{x}) + e^{y_j g(\mathbf{x})} \leq f(\mathbf{x}) + 1 < f(\mathbf{x}_j) + e^{y_j g(\mathbf{x}_j)}, \qquad (4.6)$$

precisely because $y_j g(\mathbf{x}_j) \to 0$, $\mathbf{x}_j \to \tilde{\mathbf{x}}$. But the resulting inequality (4.6) is impossible because $\mathbf{x}_j$ is a local minimizer for the auxiliary function $f(\mathbf{z}) + e^{y_j g(\mathbf{z})}$ and $\mathbf{x}$ is close to $\mathbf{x}_j$ for j large. Conclude that $\tilde{\mathbf{x}}$ is a local minimizer for (4.3).

Needless to say, under convexity conditions on f and g guaranteeing uniqueness of minimizers, the vector $\tilde{\mathbf{x}}$ can only be such minimizer. When such convexity conditions do not hold, then our algorithm may or may not converge depending on circumstances, and, in particular, on initialization vectors both for variable $\mathbf{x}$, and, possibly more importantly, for variable $\mathbf{y}$. We will have an opportunity to see these difficulties later.

This previous result ensures, regardless of any other consideration, that if the control condition $y_j g(\mathbf{x}_j) \to 0$ is monitored and satisfied during the iterative procedure, the limit vector $\tilde{\mathbf{x}}$ will be the solution sought. In practice, one can never ignore the control of the products $y_j g(\mathbf{x}_j)$ going to zero, as it is the main certificate for convergence to a true solution of our mathematical program.

Upon reflecting on the algorithm, one may be tempted to use the combination

$$f(\mathbf{x}) + y g(\mathbf{x}),$$

as long as it is coercive in $\mathbf{x}$ for positive y, instead of

$$f(\mathbf{x}) + e^{y g(\mathbf{x})},$$

in the step of the algorithm where the point of minimum is found. However, it is not reasonable to expect that the resulting algorithm would work well, because over the feasible region where $g(\mathbf{x}) < 0$, the combination $f(\mathbf{x}) + y g(\mathbf{x})$ may distort the constrained problem, and so the minimizer for $f(\mathbf{x}) + y g(\mathbf{x})$ may have little to do with the solution that is sought.

4.2.1   The Algorithm for Several Constraints

We would like to briefly describe the algorithm for a general problem like

$$\text{Minimize in } \mathbf{x} \in \mathbb{R}^N : \quad f(\mathbf{x}) \text{ subject to } \mathbf{g}(\mathbf{x}) \leq \mathbf{0}, \mathbf{h}(\mathbf{x}) = \mathbf{0},$$

where $f : \mathbb{R}^N \to \mathbb{R}$, $\mathbf{g} = (g^{(i)}) : \mathbb{R}^N \to \mathbb{R}^n$, $\mathbf{h} = (h^{(j)}) : \mathbb{R}^N \to \mathbb{R}^m$ are assumed to be smooth. We are to respect both constraints in the form of inequalities as well as in the form of equalities. Since our algorithm is specially tailored to deal with inequality restrictions, we transformed the equality restriction set $\mathbf{h}(\mathbf{x}) = \mathbf{0}$ in a double family of inequalities $\mathbf{h}(\mathbf{x}) \leq \mathbf{0}$, $-\mathbf{h}(\mathbf{x}) \leq \mathbf{0}$, and so we can focus on a general problem in which only inequalities are to be considered. We stick to this framework henceforth.

For a programming problem like

$$\text{Minimize in } \mathbf{x} \in \mathbb{R}^N : \quad f(\mathbf{x}) \quad \text{subject to} \quad \mathbf{g}(\mathbf{x}) \leq \mathbf{0}, \tag{4.7}$$

where $f : \mathbb{R}^N \to \mathbb{R}$, $\mathbf{g} = (g^{(i)}) : \mathbb{R}^N \to \mathbb{R}^n$, in which several restrictions in the form of inequalities are to be respected, the iterative algorithm would read:

1. Initialization. Set $\mathbf{y}_0 = \mathbf{1}$, $\mathbf{x}_{-1} = \mathbf{0}$ (or any other choice).
2. Iterative step until convergence. Suppose we have $\mathbf{y}_j = (y_j^{(i)})$, $\mathbf{x}_{j-1}$.
   (a) Solve (approximate) the unconstrained optimization problem

$$\text{Minimize in } \mathbf{z} \in \mathbb{R}^N : \quad f(\mathbf{z}) + \sum_{i=1}^{n} e^{y_j^{(i)} g^{(i)}(\mathbf{z})}$$

   starting from the intial guess $\mathbf{x}_{j-1}$. Let $\mathbf{x}_j$ be such minimizer.
   (b) If all products $y_j^{(i)} g^{(i)}(\mathbf{x}_j)$ vanish (are smaller than a threshold or tolerance value), stop and take $\mathbf{x}_j$ as the solution of the problem.
   (c) If not all those products vanish, update y_j by setting

$$y_{j+1}^{(i)} = e^{y_j^{(i)} g^{(i)}(\mathbf{x}_j)} y_j^{(i)}.$$

Note that, in this format, the algorithm is essentially given in pseudocode form, so that it is very easily implementable in practice, especially if one relies on a suitable subroutine to compute good approximations of unconstrained minimization problems.

The augmented objective functional above

$$f(\mathbf{z}) + \sum_{i=1}^{n} e^{y_j^{(i)} g^{(i)}(\mathbf{z})}$$

is not to be mistaken with

$$f(\mathbf{z}) + \prod_{i=1}^{n} e^{y_j^{(i)} g^{(i)}(\mathbf{z})} = f(\mathbf{z}) + \exp\left(\sum_{i=1}^{n} y_j^{(i)} g^{(i)}(\mathbf{z})\right) = f(\mathbf{z}) + e^{\mathbf{y} \cdot \mathbf{g}(\mathbf{z})}.$$

This other choice would lead to a uniform way, for all constraints at once, of updating $\mathbf{y}$ which cannot work well.

We have a convergence result quite similar to the one already proved for the one-restriction situation.

Theorem 4.2 *Let all functions involved in (4.7) be smooth. Suppose that the sequence of iterates* $\{(\mathbf{y}_j, \mathbf{g}(\mathbf{x}_j))\}$ *produced by our algorithm is such that*

$$y_j^{(i)} g^{(i)}(\mathbf{x}_j) \to 0, \quad i = 1, 2, \dots, n, \quad x_j \to \tilde{x}.$$

Then $\tilde{x}$ is a local solution of (4.7).

All comments after the proof of the corresponding theorem for the one-constraint case are also relevant here.

4.2.2 An Important Issue for the Implementation

When one comes to implement the algorithm described in the last section, it is quite important to rescale the problem appropriately before utilizing the numerical scheme. What we mean by this is that all specific numbers and parameters ocurring in the problem (both in the cost functional $f(\mathbf{x})$ and in the constraint $g(\mathbf{x})$) must be comparable to ensure the best performance of the numerical implementation. The best way of achieving this goal is to apply appropriate scaling both to the independent variable $\mathbf{x}$, to the dependent variable represented by the cost functional, and the constraints. Notice that in fact, the problem

$$\text{Minimize in } \mathbf{x} \in \mathbb{R}^N : \quad f(\mathbf{x}) \quad \text{subject to} \quad g(\mathbf{x}) \le 0$$

is equivalent to

$$\text{Minimize in } \overline{\mathbf{x}} \in \mathbb{R}^N : \quad \alpha f(\mathbf{F}(\overline{\mathbf{x}})) \quad \text{subject to} \quad \beta g(\mathbf{F}(\overline{\mathbf{x}})) \le 0,$$

where we have made the change of variables $\mathbf{x} = \mathbf{F}(\overline{\mathbf{x}})$, and α, β are arbitrary positive numbers. Most of the time a linear change of variables of the form

$$\mathbf{F}(\overline{\mathbf{x}}) = \mathbf{F}\overline{\mathbf{x}}, \quad \mathbf{F} = \text{diagonal}(\gamma_i), \quad \gamma_i > 0,$$

will suffice.

It is worth to examine the form of our main algorithm for a typical linear programming problem. Our scheme does not take advantage of linearity but it is a general-purpose procedure whose implementation is identical for all programming problems. As such, it may not be competitive compared to other ways of approximating optimal solutions that take full advantage of the particular structure or features of the problem to be solved. A main advantage of our scheme, however, is its simplicity and flexibility in practice.

Let us explore the problem

$$\text{Minimize in } \mathbf{x} \in \mathbb{R}^N \quad \mathbf{c} \cdot \mathbf{x} \quad \text{under} \quad \mathbf{A}\mathbf{x} \le \mathbf{b},$$

for vectors $\mathbf{c}$, $\mathbf{b}$, and matrix $\mathbf{A}$ of the appropriate dimension. Typical inequalities restricting the sign of variables $\mathbf{x} \ge \mathbf{0}$ are assumed to be included into the full, vector inequality $\mathbf{A}\mathbf{x} \le \mathbf{b}$, and so are feasible equalities. The augmented cost functional is

$$\mathbf{c} \cdot \mathbf{x} + \sum_i e^{y_i (\mathbf{a}^{(i)} \cdot \mathbf{x} - b_i)}, \quad \mathbf{A} = (\mathbf{a}^{(i)}), \mathbf{b} = (b_i).$$

Note how minimization with respect to $\mathbf{x}$ corresponds to a non-linear optimization problem despite the fact that the original one is a linear programming problem. As indicated, the algorithm is applied then in the same way.

4.3 Some Selected Examples

1. We consider the linear programming problem (2.5). Recall that we seek to

$$\text{Maximize in } \mathbf{x} = (x_1, x_2, x_3, x_4, x_5): \quad 12x_1 + 20x_2 + 18x_3 + 40x_4 + 10x_5$$

restricted to

$$4x_1 + 10x_2 + 8x_3 + 10x_4 + 5x_5 \le 6000,$$
$$2x_1 + x_2 + 6x_3 + 30x_4 + 3x_5 \le 4000,$$
$$3x_1 + 2x_2 + 5x_3 + 15x_4 + 5x_5 \le 5000,$$
$$x_1, x_2, x_3, x_4, x_5 \ge 0.$$

We rewrite the problem very easily in the equivalent form

$$\text{Minimize in } \mathbf{x}: \quad -(12x_1 + 20x_2 + 18x_3 + 40x_4 + 10x_5)$$

subject to

$$4x_1 + 10x_2 + 8x_3 + 10x_4 + 5x_5 \le 6000,$$
$$2x_1 + x_2 + 6x_3 + 30x_4 + 3x_5 \le 4000,$$
$$3x_1 + 2x_2 + 5x_3 + 15x_4 + 5x_5 \le 5000,$$
$$-x_1, -x_2, -x_3, -x_4, -x_5 \le 0.$$

We realize immediately that the explicit numbers involved in the ingredient $\mathbf{A}$, $\mathbf{c}$, and $\mathbf{b}$ are of two different scales. So before, we use our basic algorithm, it is important to transform the problem into a better equivalent form. To this aim, we use the new variables $\mathbf{X} = 10^{-3}\mathbf{x}$, and put the problem into the form

$$\text{Minimize in } \mathbf{X} : -1.2X_1 - 2X_2 - 1.8X_3 - 4X_4 - X_5$$

subject to

$$4X_1 + 10X_2 + 8X_3 + 10X_4 + 5X_5 \leq 6,$$
$$2X_1 + X_2 + 6X_3 + 30X_4 + 3X_5 \leq 4,$$
$$3X_1 + 2X_2 + 5X_3 + 15X_4 + 5X_5 \leq 5,$$
$$-X_1, -X_2, -X_3, -X_4, -X_5 \leq 0.$$

We introduce eight multipliers, one for each constraint,

$$\mathbf{y} = (y_1, y_2, y_3, y_4, y_5, y_6, y_7, y_8),$$

and consider the iterative scheme consisting in two main steps, and taking as initilization $\mathbf{X}_0 = \mathbf{0}$, $\mathbf{y} = \mathbf{1}$:

- Find the optimal solution (minimizer) $\mathbf{X}(\mathbf{y})$ for the unconstrained problem in $\mathbf{X}$ for given $\mathbf{y}$

$$-1.2X_1 - 2X_2 - 1.8X_3 - 4X_4 - X_5$$
$$+e^{y_1(4X_1+10X_2+8X_3+10X_4+5X_5-6)}$$
$$+e^{y_2(2X_1+X_2+6X_3+30X_4+3X_5-4)}$$
$$+e^{y_3(3X_1+2X_2+5X_3+15X_4+5X_5-5)}$$
$$+e^{-y_4X_1} + e^{-y_5X_2} + e^{-y_6X_3} + e^{-y_7X_4} + e^{-y_8X_5}.$$

- If all of the exponents in the exponentials are sufficiently small, stop and take the current $\mathbf{X}$ as the solution of the problem. If, at least one of them, is not sufficiently small, update $\mathbf{y}$ through the rule

$$y_1 = e^{y_1(4X_1+10X_2+8X_3+10X_4+5X_5-6)}y_1,$$
$$y_2 = e^{y_2(2X_1+X_2+6X_3+30X_4+3X_5-4)}y_2,$$
$$y_3 = e^{y_3(3X_1+2X_2+5X_3+15X_4+5X_5-5)}y_3,$$
$$y_4 = e^{-y_4X_1}y_4, \quad y_5 = e^{-y_5X_2}y_5, \quad y_6 = e^{-y_6X_3}y_6,$$
$$y_7 = e^{-y_7X_4}y_7, \quad y_8 = e^{-y_8X_5}y_8.$$

After several iterations, depending on the stopping criterium, one finds the following approximation:

- Optimal solution:

    $1.39986, \quad -3.167486E - 06, \quad 2.364477E - 05, \quad 3.996827E - 02, \quad 7.315134E - 05;$

- Optimal cost: -1.83982
- Multipliers:

$$6.634704E - 05, \quad 0.841745, \quad 0.680762, \quad 2.316554E - 03,$$
$$0.520671, \quad 0.274266, \quad 3.971245E - 02, \quad 4.634984E - 04;$$

- Admissibility: all these values should be nearly non-negative

$$-1.39986, \quad 3.167486E - 06, \quad -2.364477E - 05, \quad -3.996827E - 02,$$
$$-7.315134E - 05, \quad -3.499985E - 04, \quad -8.716583E - 04, \quad -0.200415;$$

- Optimality: all these values should nearly vanish

$$-9.287664E - 05, \quad 2.666215E - 06, \quad -1.609645E - 05, \quad -9.258865E - 05,$$
$$-3.808776E - 05, \quad -9.599271E - 05, \quad -3.461568E - 05, \quad -9.289209E - 05.$$

2. Let us now solve the dual problem, recast in the form

$$\text{Minimize in } \mathbf{y} = (y_1, y_2, y_3) : \quad 6y_1 + 4y_2 + 5y_3$$

subject to

$$12 - 4y_1 - 2y_2 - 3y_3 \leq 0,$$
$$20 - 10y_1 - y_2 - 2y_3 \leq 0,$$
$$18 - 8y_1 - 6y_2 - 5y_3 \leq 0,$$
$$40 - 10y_1 - 30y_2 - 15y_3 \leq 0,$$
$$10 - 5y_1 - 3y_2 - 5y_3 \leq 0,$$
$$-y_1, -y_2, -y_3 \leq 0.$$

Our algorithm with initialization $\mathbf{y} = \mathbf{1}$ yields the solution:

- Optimal solution:

$$2.80003, \quad 0.399918, \quad 3.440666E - 05$$

- Optimal cost: 18.4000.
- Multipliers:

$$1.41196, \quad 6.585274E - 06, \quad 8.121850E - 06, \quad 3.987799E - 02,$$
$$1.060633E - 05, \quad 1.975517E - 05, \quad 1.383121E - 04, \quad 0.199001;$$

- Admissibility: all these values should be nearly non-negative

$$-5.154277E - 05, \quad -8.40027, \quad -6.79991, \quad 1.651602E - 03,$$
$$-5.20007, \quad -2.80003, \quad -0.399918, \quad -3.440666E - 05;$$

- Optimality: all these values should nearly vanish

$$-7.277622E - 05, \quad -5.531805E - 05, \quad -5.522781E - 05, \quad 6.586256E - 05,$$
$$-5.515365E - 05, \quad -5.531503E - 05, \quad -5.531355E - 05, \quad -6.846970E - 06.$$

3. The building problem, Example 3.1. If we redefine variables in the way

$$x_1 = d/50, \quad x_2 = h/50, \quad x_3 = w/50,$$

then the problem is to

$$\text{Minimize in } \mathbf{x} = (x_1, x_2, x_3): \quad x_1 x_3^2$$

subject to

$$\frac{4 \times 3.5}{25 \times 1.618} - x_3^2(x_1 + x_2) \leq 0, \quad x_3 - \frac{1}{1.618} \leq 0,$$
$$2 \times 2.618 x_2 x_3 + 1.618 x_3^2 - 0.9 \leq 0, \quad -x_1, -x_2, -x_3 \leq 0.$$

This is an interesting example to argue about the importance of having the hypotheses of Theorem 4.2. If one would start with the initial guess $y_i = 1$. for all components i, the iterative procedure may not produce anything meaningful, if it at all converges. One has to scale appropriately these multipliers by setting the initial values to

$$y_1 = 1., \quad y_2 = 1., \quad y_3 = 1., \quad y_4 = 50., \quad y_5 = 50., \quad y_6 = 50.$$

Note that these big values correspond to the non-negativeness of the three variables x_1, x_2, x_3 which have been rescaled. Then the iterative procedure yields the following correct optimal data set:

- Optimal solution:
$$1.61429, \quad 0.267958, \quad 0.428771$$

- Optimal cost: 0.296779.
- Multipliers:

$$1.00227, \quad 4.955553E - 04, \quad 8.199302E - 02,$$
$$9.368889E - 21, \quad 1.879989E - 10, \quad 2.292688E - 04;$$

- Admissibility: all these values should be nearly non-negative

$$5.796552E - 05, \quad -0.189229, \quad -9.602308E - 04,$$
$$-1.61429, \quad -0.267958, \quad -0.428771;$$

- Optimality: all these values should nearly vanish

$$5.809714E - 05, \quad -9.377327E - 05, \quad -7.873223E - 05,$$
$$-1.512412E - 20, \quad -5.037585E - 11, \quad -9.830387E - 05.$$

These optimal values correspond to the real quantities

$$d = 80.7145, \quad h = 13.3979, \quad w = 21.4385, \quad l = 34.6875,$$

and the amount of land to be removed 60023.55 cubic units.

4. We now turn to example HS071 from [19]. The specific problem is to

$$\text{Minimize in}(x_1, x_2, x_3, x_4): \quad x_1 x_4(x_1 + x_2 + x_3) + x_3$$

under the constraints

$$x_1 x_2 x_3 x_4 \geq 25, \quad x_1^2 + x_2^2 + x_3^2 + x_4^2 = 40,$$
$$1 \leq x_1, x_2, x_3, x_4 \leq 5.$$

For this situation, we take as initial guess $x_i = 2.$ for all i, and

$$y_1 = y_2 = y_3 = y_4 = 15., \quad y_j = 1. \text{ for } j \geq 5.$$

The approximated solution obtained is

- Optimal solution:

$$0.999922, \quad 4.68520, \quad 3.89530, \quad 1.36996$$

- Optimal cost: 17.0191.
- Multipliers:

$$1.28594, \quad 2.679033E - 28, \quad 3.304557E - 05, \quad 2.561671E - 04, \quad 2.375436E - 05,$$
$$3.163394E - 04, \quad 8.444802E - 05, \quad 2.618186E - 05, \quad 53.6916, \quad 1401.51;$$

- Admissibility: all these values should be nearly non-negative

$$7.772446E - 05, \quad -3.68520, \quad -2.89530, \quad -0.369958, \quad -4.00008,$$
$$-0.314795, \quad -1.10470, \quad -3.63004, \quad -1.831055E - 06, \quad 1.302588E - 08;$$

- Optimality: all these values should nearly vanish

$$9.994920E - 05, \quad -9.872785E - 28, \quad -9.567687E - 05,$$
$$-9.477117E - 05, \quad -9.501927E - 05, \quad -9.958223E - 05,$$
$$-9.328964E - 05, \quad -9.504125E - 05, \quad -9.831226E - 05, \quad 1.825596E - 05.$$

5. Finally, we examine example HS114 of that same reference. This is a much
   more involved and demanding example as the number of variables, and non-
   linear constraints is higher. We refer to [19] for details. The optimal value we
   have found, by using the initialization

$$\mathbf{y} = 1, \quad x_1 = 1745., x_2 = 12000., x_3 = 110., x_4 = 2048.,$$
$$x_5 = 1974., x_6 = 89.2, x_7 = 92.8, x_8 = 8., x_9 = 3.6, x_{10} = 145,$$

is -2400.54 attained at the optimal point

$$1705.08, \quad 12000.3, \quad 9.453460E - 02, \quad 3015.26, \quad 1973.51,$$
$$93.0137, \quad 94.9987, \quad 8.19182, \quad 1.72742, \quad 152.737.$$

4.4 Final Remarks

Approximation techniques for optimization problems is a fundamental field of mathematical programming with a relevance well beyond Mathematics. It is about real
life. No wonder that software commercial packages of various sorts and of various
features are available in the market. The difficulties with large scale optimization
problems require rather fine strategies and ideas that deserve full consideration. Real
life problems demand real experts in the approximation arena. Our aim in this chapter
has been to introduce some simple techniques that can be helpful at least for some
typical academic examples. Having an affordable tool to compute approximations to
optimal solutions to problems can also be very motivating to students, for otherwise
discouragement may pop up as they realize that essentially none of the interesting
problems can be solved or approximated without these approximation tools.

4.5 Exercises

4.5.1 Exercises to Support the Main Concepts

1. For $y > 0$, consider the function

$$L(x_1, x_2, x_3, y) = x_1 - 2x_2 - x_3 + y(x_1^2 + x_2^2 + x_3^2 - 1).$$

If we put $\mathbf{x} = (x_1, x_2, x_3)$, $g(\mathbf{x}) = x_1^2 + x_2^2 + x_3^2 - 1$, and $\mathbf{x}_y$ is the unique point
of minimum of L with respect to $\mathbf{x}$ for each fixed y, find explicitly the function
$G(y) = ye^{yg(\mathbf{x}_y)}$, and represent its graph. Is there a point y where $G(y) = y$?
What is special about the vector $\mathbf{x}_y$ for such a value of y?

2. Consider the problem

$$\text{Minimize in } \mathbf{x} \in \mathbb{R}^N : \quad f(\mathbf{x})$$

over the set where $g(\mathbf{x}) \leq 0$, where both f, and g are smooth, strictly convex functions. Consider the master function

$$L(\mathbf{x}, y) = f(\mathbf{x}) + e^{yg(\mathbf{x})}.$$

We know that for $y \geq 0$, $L(\mathbf{x}, y)$ is strictly convex in $\mathbf{x}$. Let $\mathbf{x}_y$ be the unique point of minimum. Define $G : \mathbb{R}^+ \to \mathbb{R}^+$ through

$$G(y) = ye^{yg(\mathbf{x}_y)}.$$

The purpose of this problem is to study some properties of this function $G(y)$ for $y > 0$.

(a) Because $\mathbf{x}(y)$ is a point of minimum for $L(\mathbf{x}, y)$ with respect to $\mathbf{x}$, show that

$$\nabla f(\mathbf{x}) + e^{yg(\mathbf{x})} y \nabla g(\mathbf{x}) = \mathbf{0},$$
$$\nabla^2 f(\mathbf{x}) + e^{yg(\mathbf{x})} y^2 \nabla g(\mathbf{x}) \otimes \nabla g(\mathbf{x}) + e^{yg(\mathbf{x})} y \nabla^2 g(\mathbf{x}) \geq 0,$$

as symmetric matrices.

(b) Conclude that

$$(1 + yg(\mathbf{x})) \frac{d}{dy} g(\mathbf{x}(y)) \leq 0$$

for all y.

(c) Show that

$$G'(y) \frac{d}{dy} g(\mathbf{x}(y)) \leq 0$$

for all y.

From these two properties, one can show that $G(y)$ admits a fixed point $\overline{y} \geq 0$ such that $\mathbf{x}(\overline{y})$ is precisely the solution of the corresponding constrained problem, though this a much more advanced task.

3. For the LP problem

$$\text{Minimize in } (x_1, x_2) : \quad x_1 - 2x_2$$

under

$$x_1 \geq 0, \quad x_1 - x_2 \leq -1, \quad -x_1 + 3x_2 \geq 0,$$

write in detail the numerical algorithm to approximate its optimal solution, the non-linear system to find the x's variables in terms of the variable y's, and the update rule for the variables y's.

4. For a vector function $\mathbf{G}(\mathbf{y})$ for which we would like to approximate a fixed point $\mathbf{G}(\mathbf{y}) = \mathbf{y}$, we can use Newton's method in the following way to produce an iterative approximation procedure. Given iterate $\mathbf{y}_j$, find the next one $\mathbf{y}_{j+1}$ by demanding, ideally, that

$$\mathbf{G}(\mathbf{y}_{j+1}) = \mathbf{0}.$$

If we use a Taylor development of first-order of $\mathbf{G}$ at the base point $\mathbf{y}_j$, we would ask for

$$\mathbf{G}(\mathbf{y}_j) + \nabla\mathbf{G}(\mathbf{y}_j) \cdot (\mathbf{y}_{j+1} - \mathbf{y}_j) = \mathbf{0},$$

and so take

$$\mathbf{y}_{j+1} = \mathbf{y}_j + \nabla\mathbf{G}(\mathbf{y}_j)^{-1}\mathbf{G}(\mathbf{y}_j).$$

Apply this method to the function $G(y)$ associated with the optimization problem corresponding to the master function

$$L(x_1, x_2, x_3, y) = x_1 - 2x_2 - x_3 + y(x_1^2 + x_2^2 + x_3^2 - 1)$$

of the first exercise.

5. Conjugate gradient method ([8, 16, 25]). Consider the quadratic function

$$q(\mathbf{x}) = \frac{1}{2}\mathbf{x}^*\mathbf{A}\mathbf{x} + \mathbf{b}\mathbf{x} + c$$

for a symmetric, positive definite, $N \times N$-matrix $\mathbf{A}$, a constant vector $\mathbf{b}$, and a number c. $\mathbf{x} \in \mathbb{R}^N$.

(a) Consider the inner product in $\mathbb{R}^N$ given by

$$\langle \mathbf{x}, \mathbf{y} \rangle = \mathbf{x}^*\mathbf{A}\mathbf{y}.$$

Take a basis $\{\mathbf{e}_i\}$ of $\mathbb{R}^N$, and use the standard Gram-Schmidt orthogonalization process to find a orthogonal basis from the selected one.

(b) Suppose $\{\mathbf{e}_i\}$ is already an orthogonal basis so that

$$\langle \mathbf{e}_i, \mathbf{e}_j \rangle = \delta_{ij}.$$

Write a process in N steps to find the global minimum of q, by using $\mathbf{e}_i$ as successive search directions.

(c) Prove that indeed the result of the previous scheme leads to the minimum of q.

6. For the Rosenberg function[1]

---

[1] From [19].

$$f(x_1, x_2) = 100(x_2 - x_1^2)^2 + (1 - x_1)^2,$$

examined the level curves

$$f(x_1, x_2) = c$$

for various values of the constant c with the help of an appropriate software package. What is the minimum value, and the point where it is attained? Compute the negative of the gradient of f at $(0, 0)$. Does this vector point towards the minimum value of f?

7. The solution of the linear system $\mathbf{Ax} = \mathbf{b}$ can be approximated by minimizing the function

$$E(\mathbf{x}) = \frac{1}{2}|\mathbf{Ax} - \mathbf{b}|^2,$$

or, equivalently, by minimizing

$$\tilde{E}(\mathbf{x}) = \frac{1}{2}\mathbf{x}^*\mathbf{A}^*\mathbf{Ax} - \mathbf{A}^*\mathbf{bx}.$$

Write explicitly the main iterative step to solve the linear system

$$x_1 - x_2 + 2x_3 - 2x_4 = 3, \quad 2x_1 + x_2 + 2x_3 - x_4 = 6,$$
$$x_1 + x_3 + x_3 + x_4 = 6, \quad 3x_1 - x_2 - 3x_3 + x_4 = 0,$$

through a descent strategy.

4.5.2 Computer Exercises

1. Design a subroutine to approximate the minimum of a strictly convex function $f(\mathbf{x})$ subjected to no constraint $\mathbf{x} \in \mathbb{R}^N$. Test it with the following examples:

    (a) $f(\mathbf{x}) = (x_1 - 1)^2 + (x_2 + 1)^4$;
    (b) $f(\mathbf{x}) = \mathbf{a} \cdot \mathbf{x} + e^{\mathbf{b} \cdot \mathbf{x}}$ with $\mathbf{a} = (1, 1, -1)$, $\mathbf{b} = (-1, 1, 1)$;
    (c) $f(\mathbf{x}) = \mathbf{a} \cdot \mathbf{x} + e^{\mathbf{b} \cdot \mathbf{x}} + e^{\mathbf{c} \cdot \mathbf{x}}$ with $\mathbf{a} = (1, 1, -1)$, $\mathbf{b} = (-1, 1, 1)$, $\mathbf{c} = (1, -1, 1)$.

2. Modify the previous subroutine to accomodate constraints of the form $g(\mathbf{x}) \leq 0$ according to the numerical strategy described in this chapter.

3. Approximate the solution of the problem[2]

$$\text{Maximize in } (x_1, x_2): \quad x_1 + x_2$$

---

[2]From [20].

under

$$2\alpha x_1 + x_2 \le \alpha^2, \quad \alpha = \{0, 0.1, 0.2, 0.3, 0.4, 0.5, 0.6, 0.7, 0.8, 0.9, 1.\}.$$

4. Approximate the minimum of the problem[3]

$$f(x_1, x_2, x_3, x_4, x_5) = (x_1 - x_2)^2 + (x_2 - x_3)^2 + (x_3 - x_4)^2 + (x_4 - x_5)^2$$

under the constraints

$$x_1 + 2x_2 + 3x_3 = 6, \quad x_2 + 2x_3 + 3x_4 = 6, \quad x_3 + 2x_4 + 3x_5 = 6.$$

The solution is (1, 1, 1, 1, 1).
5. Approximate the minimum of[4]

$$f(x_1, x_2, x_3) = (x_1 - 1)^2 + (x_1 - x_2)^2 + (x_2 - x_3)^4$$

subjected to

$$x_1(1 + x_2^2) + x_3^4 = 4 + \frac{3}{\sqrt{2}}.$$

Use $(2, -10, 10)$ as a starting point.
6. Approximate the minimum value of the function

$$f(x_1, x_2, x_3) = x_1^2 + x_2^2 + x_3^2$$

subject to $x_1 x_2 + x_1 x_3 + x_2 x_3 = 1$.
7. Explore the minimum of the polynomial[5]

$$P(x_1, x_2) = 2x_1^4 + 2x_1^3 x_2 - x_1^2 x_2^2 + 5x_2^4.$$

8. Examine the minimum value for the two polynomials[6]

$$p(x_1, x_2) = x_1^2 x_2^2 (x_1^2 + x_2^2 - 1) + 1,$$
$$p(x_1, x_2, x_3) = x_1^2 x_2^2 (x_1^2 + x_2^2 - 3x_3^2) + 6x_3^6.$$

9. Study, either numerically or analytically, the problem[7]

$$\text{Minimize in}(x_1, x_2): \quad x_1^3 - x_1^2 + 2x_1 x_2 - x_2^2 + x_2^3$$

---

[3]From [19].
[4]From [19].
[5]From [21].
[6]From [21].
[7]From [21].

in the subset determined by the inequalities

$$x_1 \geq 0, \quad x_2 \geq 0, \quad x_1 + x_2 \geq 1.$$

10. Show that the smallest eigenvalue of the symmetric matrix $\mathbf{A}$ is the minimum of the problem

$$\text{Minimize in } \mathbf{x}: \quad \frac{1}{2}\mathbf{x}^*\mathbf{A}\mathbf{x}$$

under $|\mathbf{x}|^2 = 1$. Find an approximation of the smallest eigenvalue of

$$\mathbf{A} = \begin{pmatrix} 1. & 1.1 & 1.11 \\ 1.1 & 1. & 2.1 \\ 1.11 & 2.1 & 1. \end{pmatrix}.$$

4.5.3 Case Studies

We propose three typical situations of linear programming problems.

1. **A transportation case.** Let us consider the following transport problem[8] according to the figure, with $m = 3$ starting points and $n = 3$ destinations, and

$$u_1 = 2, u_2 = 3, u_3 = 4; v_1 = 5, v_2 = 2, v_3 = 2$$

The matrix of costs is

$$\mathbf{c} = \begin{pmatrix} 1 & 2 & 3 \\ 2 & 1 & 2 \\ 3 & 2 & 1 \end{pmatrix}$$

where each entry c_{ij} is the cost of the $i - j$ route (x_{ij} has a unitary cost c_{ij}). Approximate the optimal solution for delivery minimizing costs.

2. **A diet problem.** Table 4.1 provides the basic nutrition information for a desired diet.[9] If unitary prices of the five foods are

$$c_1 = 1, c_2 = 0.5, c_3 = 2, c_4 = 1.2, c_5 = 3,$$

approximate one such diet of minimal price, guaranteeing that the minimum required amount of each nutrient is met.

3. **The portfolio problem.** A certain investor owns shares of several stocks: he has b_i shares of $A_i, i = 1, 2, ..., m$.[10] Current prices are v_i. Suppose we have information

---

[8]From [10].
[9]From [10].
[10]From [10].

Table 4.1 Nutrient content of five foods: (DN) digestible nutrients, (DP) digestible proteins, (Ca) Calcium, y (Ph) Phosphorus

Nutrient	Required amount	Corn A	Oat	Corn B	Bran	Linseed
DN	74.2	78.6	70.1	80.1	67.2	77.0
DP	14.7	6.50	9.40	8.80	13.7	30.4
Ca	0.14	0.02	0.09	0.03	0.14	0.41
Ph	0.55	0.27	0.34	0.30	1.29	0.86

on dividends that will be paid at the end of the current year, and final prices of each title: A_i will pay d_i and will have a new price w_i.

Our goal is to adjust our portfolio, i.e. the number of shares for each title, to maximizes dividends. The unknowns are x_i, the change in the number of shares that we already have. Since at the beginning we owned b_i, after the adjustment we will have $b_i + x_i$, respectively.

The main ingredients of the problem are:

(a) **Data**

   m: the number of titles involved;

   b_i: the number of current share for stock i;

   v_i: price per share for stock k;

   d_i: dividend that will be paid at the end of the current year for title i;

   w_i: new price of stock i;

   r: minimal percentage of the total current value of the portfolio that cannot be exceeded in the adjustment;

   s: minimal percentage of the total current price that cannot be overcome by the total future price of the portfolio, aiming at facing inflation.

(b) **Variables**

   x_i: the change in the number of shares for each title i.

(c) **Constraints** Though not yet expressed, we ought to ensure certain working conditions for a well-balanced portfolio. The number of shares can never be negative:

$$x_i \geq -b_i$$

An excessive concentration of attention on any particular stock ought to be avoided; this condition can be implemented by demanding that the capital associated with any particular stock after adjustment be at least a certain percentage r of the total current capital. This means

$$r\left(\sum_i v_i(b_i + x_i)\right) \leq v_j(b_j + x_j) \quad \text{for all } j.$$

The global value of the portfolio cannot change in the adjustment, as we assume no additional investment

$$\sum_i v_i x_i = 0.$$

To face inflation, the total value of the portfolio in the future must be, at least, a certain fraction s greater that the invested current capital

$$\sum_i w_i (b_i + x_i) \geq (1 + s) \sum_i v_i b_i.$$

(d) **Objective to be maximize** Our goal is to maximize dividends in the adjustment

$$Z = \sum_i d_i (b_i + x_i),$$

but preserving all of the conditions described above.

After stating the problem in precise and compact terms, solve the particular case in which we have shares of three stocks, 75 of A, 100 of B, and 35 of C, with prices US $20, US $20 and US $100, respectively. In addition, the following information is at our disposal: A will not pay dividends and will have a new price US $18, B will pay US $3 per share and the new price will be US $23, and C will pay US $5 per share with a new price US $102. Percentages r and s are, respectively, 0.25 and 0.03.

References

1. N. Andrei, *Nonlinear Optimization Applications using the GAMS Technology, Springer Optimization and Its Applications*, vol. 81 (Springer, New York, 2013)
2. R.K. Arora, *Optimization: Algorithms and Applications* (CRC Press, Boca Raton, FL, 2015)
3. P. Bangert, *Optimization for Industrial Problems* (Springer, Heidelberg, 2012)
4. A. Beck, *Introduction to Nonlinear Optimization. Theory, Algorithms, and Applications with MATLAB*, MOS-SIAM Series on Optimization. Society for Industrial and Applied Mathematics (SIAM), Mathematical Optimization Society, vol. 19 (Philadelphia, PA, 2014)
5. M.A. Bhatti, *Practical Optimization Methods with Mathematica Applications* (Springer, New York, 2000)
6. M. Bierlaire, *Optimization: Principles and Algorithms* (EPFL Press, Lausanne; distributed by CRC Press, Boca Raton, FL, 2015)
7. E.G. Birgin, J.M. Martnez, *Practical Augmented Lagrangian Methods for Constrained Optimization*, Fundamentals of Algorithms. Society for Industrial and Applied Mathematics (SIAM), vol. 10 (Philadelphia, PA, 2014)
8. S. Butenko, P.M. Pardalos, *Numerical Methods and Optimization. An Introduction*, By Chapman and Hall edn., CRC Numerical Analysis and Scientific Computing Series (CRC Press, Boca Raton, 2014)
9. Ch.L. Byrne, *A First Course in Optimization* (CRC Press, Boca Raton, FL, 2015)
10. E. Castillo, A. Conejo, P. Pedregal, R. García, N. Alguacil, *Building and Solving Mathematical Programming Models in Engineering and Science*, Pure and Applied Mathematics (Wiley, New York, 2002)
11. P.W. Christensen, A. Klarbring, *An Introduction to Structural Optimization*, Solid Mechanics and its Applications, vol. 153 (Springer, New York, 2009)

12. http://conopt.com/
13. A. Dhara, J. Dutta, *Optimality Conditions in Convex Optimization. A Finite-dimensional View*, with a foreword by Stephan Dempe (CRC Press, Boca Raton, FL, 2012)
14. Decision tree for optimization software: http://plato.asu.edu/guide.html
15. U. Faigle, W. Kern, G. Still, *Algorithmic Principles of Mathematical Programming*, Kluwer Texts in the Mathematical Sciences, vol. 24 (Kluwer Academic Publishers, Dordrecht, 2002)
16. A.V. Fiacco, G.P. McCormick, *Nonlinear Programming. Sequential Unconstrained Minimization Techniques*, SIAM Classics in Applied Mathematics, vol. 4 (Philadelphia)
17. https://gams.com/
18. J.B. Hiriart-Urruty, C. Lemarchal, *Convex Analysis and Minimization Algorithms. I. Fundamentals*, Grundlehren der Mathematischen Wissenschaften [Fundamental Principles of Mathematical Sciences], vol. 305 (Springer-Verlag, Berlin, 1993)
19. W. Hock, K. Schittkowski, *Test Examples for Nonlinear Programming Codes*, Lecture Notes in Economics and Mathematical Systems, vol. 187 (Springer, Berlin, 1981)
20. https://en.wikipedia.org/wiki/Interior-point-method
21. J.B. Laserre, *Moments, Positive Polynomials and Their Applications*, vol. 1 (Imperial College Press Optimization Series, London, 2010)
22. I. Maros, *Computational Techniques of the Simplex Method*, with a foreword by András Prékopa. International Series in Operations Research and Management Science, vol. 61 (Kluwer Academic Publishers, Boston, 2003)
23. www.mathworks.com
24. P.B. Morgan, *An Explanation of Constrained Optimization for Economists* (University of Toronto Press, Toronto, ON, 2015)
25. J. Nocedal, S.J. Wright, *Numerical Optimization, Springer Series in Operations Research and Financial Engineering* (Springer, Berlin, 1999)
26. P. Pedregal, *Introduction to Optimization*, Texts Appl. Math., vol. 46 (Springer, 2003)
27. G. Sierksma, Y. Zwols, *Linear and Integer Optimization. Theory and Practice*, 3rd edn. Advances in Applied Mathematics (CRC Press, Boca Raton, FL, 2015)
28. S.A-H. Soliman, A-A.H. Mantawy, *Modern Optimization Techniques with Applications in Electric Power Systems*, Energy Systems (Springer, New York, 2012)
29. A. Schrijver, *Theory of Linear and Integer Programming* (Wiley, New York, 1999)
30. A. Takayama, *Mathematical Economics*, 2nd edn. (Cambridge University Press, 1985)
31. B. Vitoriano, *Programación matemática: Modelos de optimización* (in spanish), personal manuscript, U. Complutense (2010)
32. http://web.mit.edu/15.053/www/AMP-Chapter-11.pdf
33. H.P. Williams, *Model Building in Mathematical Programming*, 5th edn. (John Wiley and Sons, Chichester, 2013)
34. A.J. Zaslavski, *Numerical Optimization with Computational Errors*, Springer Optimization and Its Applications, vol. 108 (Springer, [Cham], 2016)

Part II
Variational Problems

Chapter 5
Basic Theory for Variational Problems

We start in this chapter a rather different, but equally relevant, part of Optimization. There is a striking main initial difference, with respect to mathematical programs, that is a clear indication that methods and techniques will have to be quite different for these problems. We mean that we will be dealing with finding extremals for functionals rather than functions. The term is used to stress the fact that competing objects are no longer vectors (of a certain finite dimension), but rather functions belonging to infinite-dimensional spaces. So, among a bunch of possibilities represented by functions with certain properties, we seek the one(s) that provide the minimum possible value of a functional defined on them. The analysis of this kind of problems involves many subtle issues which we will not cover here. This chapter aims at being a succinct introduction to this fundamental subject that will endow readers with a quick overview of the nature of these problems.

5.1 Some Motivating Examples

As usual, there is no better way to understand the distinct features of a family of problems that to examine some well-chosen cases. The first two we are going to examine are among the classical examples in every introductory textbook to the Calculus of Variations. The third one comes from Mathematical Economics.

Example 5.1 The brachistochrone. Given two points in a plane at different height, find the profile of the curve through which a unit mass under the action of gravity, and without sliping, employs the shortest time to go from the highest point to the lowest. Let $u(x)$ be one such feasible profile so that $u(0) = 0$, $u(a) = A$ with $a, A > 0$, so that we consider curves joining the two points $(0, 0)$, (a, A) in the plane, Fig. 5.1. It is easy to realize that we can restrict attention to graphs of functions like the one represented by $u(x)$ because curves joining the two points which are not

© Springer International Publishing AG 2017

P. Pedregal, *Optimization and Approximation*, UNITEXT 108,

DOI 10.1007/978-3-319-64843-9_5

Fig. 5.1 The
brachistochrone

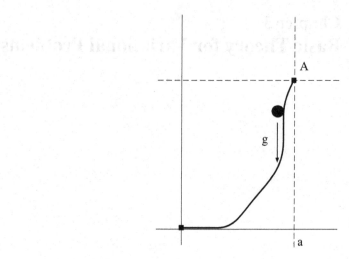

graphs cannot provide a minimum transit time under the given conditions. We need
to express, for such a function $u(x)$, the time spent by the unit mass in going from
the highest point to the lowest. From elementary kinematics, we know that

$$dt = \frac{ds}{\sqrt{2gh}}, \quad T = \int dt = \int \frac{ds}{\sqrt{2gh}},$$

where t is time, s is arc-length (distance), g is gravity, and h is height. We also know
that $ds = \sqrt{1 + u'(x)^2}\, dx$, and can identify h with $u(x)$. Altogether we find that the
total transit time T is

$$T = \frac{1}{\sqrt{2g}} \int_0^a \frac{\sqrt{1 + u'(x)^2}}{\sqrt{u(x)}}\, dx.$$

An additional simplification can be implemented if we place the X-axis vertically,
instead of horizontally, without changing the setup of the problem. Then we would
have that

$$T = \frac{1}{\sqrt{2g}} \int_0^A \frac{\sqrt{1 + u'(x)^2}}{\sqrt{x}}\, dx, \tag{5.1}$$

and the problem we would like to solve is

$$\text{Minimize in } u(x): \quad \int_0^A \frac{\sqrt{1 + u'(x)^2}}{\sqrt{x}}\, dx$$

subject to $u(0) = 0$, $u(A) = a$. Notice that this new function u is the inverse of the
old one.

Fig. 5.2 Three feasible
profiles

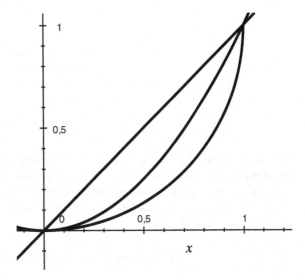

To stress the meaning of our problem, suppose, for the sake of definiteness, that we
take $a = A = 1$. In this particular situation (Fig. 5.2), we have three easy possibilities
of a function passing through the two points $(0, 0)$, $(1, 1)$, namely

$$u_1(x) = x, \quad u_2(x) = x^2, \quad u_3(x) = 1 - \sqrt{1 - x^2}.$$

Let us compare the transit time for the three, and decide which of the three yields
a smaller value. According to (5.1), we have to compute (approximate) the value of
the three integrals

$$T_1 = \int_0^1 \frac{\sqrt{2}}{\sqrt{x}} \, dx,$$

$$T_2 = \int_0^1 \frac{\sqrt{1 + 4x^2}}{\sqrt{x}} \, dx,$$

$$T_3 = \int_0^1 \frac{1}{\sqrt{x(1 - x^2)}} \, dx.$$

Note that the three are improper integrals because the three integrands do have, at
least, an asymptote at zero. Yet the value of the three integrals is finite. The smaller
of the three is the one corresponding to the parabola u_2. But still the important issue
is to find the best of all such profiles.

Example 5.2 The hanging cable. This time we have a uniform, homogeneous cable
of total length L that is to be suspended from its two end-points between two points
at the same height, and separated a distance H apart, Fig. 5.3. We will necessarily

Fig. 5.3 The hanging cable

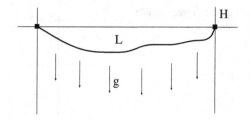

have $L > H$. We would like to figure out the shape that the hanging cable will adopt under the action of its own weight. If we assume that such a profile will be the result of minimizing the potential energy associated with any such admissible profile represented by the graph of a function $u(x)$, then we know that potential energy is proportional to height, and so

$$dP = u(x)\,ds, \quad ds = \sqrt{1 + u'(x)^2}\,dx.$$

The full potential energy contained in such feasible profile will then be

$$P = \int_0^H u(x)\sqrt{1 + u'(x)^2}\,dx.$$

Constraints now should account for the fact that the cable has total length L, in addition to demanding $u(0) = 0$, $u(H) = 0$. The constraint on the length reads

$$L = \int_0^H \sqrt{1 + u'(x)^2}\,dx,$$

coming from integrating in the full interval $[0, H]$ arc-length ds. We then seek to find the optimal shape corresponding to the problem

$$\text{Minimize in } u(x): \quad \int_0^H u(x)\sqrt{1 + u'(u)^2}\,dx$$

under the constraints

$$u(0) = u(H) = 0, \quad L = \int_0^H \sqrt{1 + u'(x)^2}\,dx.$$

Example 5.3 An optimal growth problem for an aggregate economy ([26]). An aggregate model of economic growth can be described by the (non-linear) differential equation

$$k'(t) = f(k(t)) - \lambda k(t) - x(t), \quad t \in (0, T), \tag{5.2}$$

where $k(t)$ stands for the ratio capital/labor, and $x(t)$ is consumption. The function f is given in the model, and λ is a parameter. T is the horizon considered, and k must comply with initial and terminal conditions $k(0) = k_0$, $k(T) = k_T$ which are non-negative. The objective, in this model, is to find the optimal $k(t)$ so as to maximize the global satisfaction of a representative individual in this model measured by the integral

$$I(k) = \int_0^T U(x(t))e^{-\rho t}\, dt.$$

Here $U(x)$ is the utility function for which we have $U'(x) > 0$, $U''(x) < 0$ for $x \geq 0$. It is therefore a concave, increasing function of consumption. The parameter ρ accounts for the discounting of the future. If we solve for $x(t)$ in (5.2), and plug it in $I(k)$, we arrive at the variational problem

$$\text{Maximize in } k(t): \quad I(k) = \int_0^T U(f(k(t)) - \lambda k(t) - k'(t))e^{-\rho t}\, dt$$

subject to $k(0) = k_0$, $k(T) = k_T$. The integrand for this problem is

$$F(t, u, p) = U(f(u) - \lambda u - p)e^{-\rho t}.$$

Through the study of these three examples, we see that the main ingredient of a variational problem is the functional I through which we can compare, and decide the better of various possibilities represented by functions $u(x)$. To homogenize notation, we will be using the independent variable t instead of x, as we will reserve x, y, z, for components of vectors. Most of the time such a functional will adopt the form

$$I(u) = \int_0^T F(t, u(t), u'(t))\, dt$$

where the integrand

$$F(t, u, p) : [0, T] \times \mathbb{R} \times \mathbb{R} \to \mathbb{R}$$

is a function of three variables (t, u, p). The variable p is the place occupied by the derivative u' when computing the cost of the possibility represented by u. As a matter of fact, with the same effort, we can consider a more general framework in which

$$F(t, \mathbf{u}, \mathbf{p}) : [0, T] \times \mathbb{R}^N \times \mathbb{R}^N \to \mathbb{R}$$

and competing object are now identified with vector-valued maps $\mathbf{u}(t) : [0, T] \to \mathbb{R}^N$ whose derivative is $\mathbf{u}'(t) : [0, T] \to \mathbb{R}^N$. These can be thought of as curves in N-dimensional space.

Another main ingredient in a variational problem, as in every mathematical program, is the set of restrictions. Quite often in variational problems constraints reduce to demanding that competing functions comply with prescribed values at both end-points, or at one of them. But other constraints are also possible either in the form of another integral (like in the hanging cable problem) to be respected; or in the form of pointwise constraints for every single value of the independent variable $t \in [0, T]$. Since these last constraints are dealt with in a different way compared to end-point conditions, we will first focus on the former, and then move on to the latter.

5.2   Existence of Optimal Solutions

We will assume that a certain fixed integrand

$$F : [0, T] \times \mathbb{R}^N \times \mathbb{R}^N \to \mathbb{R}$$

is given to us, and for smooth curves $\mathbf{u} : [0, T] \to \mathbb{R}^N$, we calculate costs through the integrals

$$I(\mathbf{u}) = \int_0^T F(t, \mathbf{u}(t), \mathbf{u}'(t)) \, dt.$$

We will also assume that restrictions on participating curves are just expressed in the form $\mathbf{u}(0) = \mathbf{u}_0$, $\mathbf{u}(T) = \mathbf{u}_T$ where both vectors $\mathbf{u}_0, \mathbf{u}_T \in \mathbb{R}^N$ are known. One such feasible curve is

$$\mathbf{u}(t) = \frac{t}{T}\mathbf{u}_T + \frac{(T - t)}{T}\mathbf{u}_0, \quad t \in [0, T],$$

with cost

$$\int_0^T F\left(t, \frac{t}{T}\mathbf{u}_T + \frac{(T - t)}{T}\mathbf{u}_0, \frac{\mathbf{u}_T - \mathbf{u}_0}{T}\right) dt.$$

The first important issue is to have a practical criterium to ensure, at least for a rather big family of problems, that there is an optimal path $\mathbf{u}$ among all those admissible. In the case of a mathematical program in which we look for a minimum, this was easy through the Weierstrass' criterium. It amounted to checking, under continuity of the cost function, that either the feasible set of vectors was bounded; or else, that the cost function was coercive, at least restricted to the feasible set. In the situation in which we face a variational problem, the discussion is not so elementary, and this is essentially due to the fact that we are now working in an infinite-dimensional environment. The continuity of the integrand F for the functional I does not suffice, in general, to have the existence of a minimal curve, even if the functional is coercive. A new property of the functional I, in addition to the coercivity, is necessary: the so-called weak lower semicontinuity. This is beyond the scope of this short introduction to the subject. What is quite remarkable is that the property of the integrand F that ensures that funny weak lower semicontinuity is again the main fundamental ingredient for mathematical programs: convexity.

Since the scope of this text does not pretend to be that of a complete mathematical treatise on the Calculus of Variations, we will take for granted the following results. We are more interested in understanding and manipulating optimality conditions, as well as in implementing numerical approximations of optimal profiles.

Theorem 5.1 *Let the integrand* $F(t, \mathbf{u}, \mathbf{p}) : [0, T] \times \mathbb{R}^N \times \mathbb{R}^N \to \mathbb{R}$ *be smooth with respect to all its variables. Suppose further that*

1. coercivity:
$$C(|\mathbf{p}|^r - 1) \leq F(t, \mathbf{u}, \mathbf{p}), \quad C > 0, r > 1$$

for all triplets $(t, \mathbf{u}, \mathbf{p})$;
2. convexity: the function $\mathbf{p} \mapsto F(t, \mathbf{u}, \mathbf{p})$ *is convex for every individual choice* $(t, \mathbf{u}) \in [0, T] \times \mathbb{R}^N$.

Then the variational problem associated with this integrand F, *and subjected to end-point conditions, admits an optimal solution* $\mathbf{u}$.

We have not said anything explicit about the properties of this optimal path $\mathbf{u}$, but we have done that on purpose. The selection of appropriate functional spaces where solutions are sought and found is of paramount importance. Again our aim here is to convey the main ideas and techniques, and are not interested in spending any effort on those issues which are, most definitely, part of a second round on this area.

Theorem 5.1 plays a similar role for variational problems as Weierestrass' criterium for mathematical programs. The failure of convexity on the $\mathbf{p}$-variable, and how it could affect the lack of optimal solutions, is, however, different from the finite dimensional situation. Even if the coercivity condition does hold, one could find real troubles in finding optimal solutions in typical functional spaces. But again these issues greatly exceeds the scope of this text. In our analysis of examples, we will be mainly concerned with the convexity on the variable $\mathbf{p}$, and ignore all other requirements. From a strictly approximation viewpoint, this convexity is the condition that ensures that the corresponding variational problem admits (global) optimal solutions. However, there might also be local solutions much in the same way as with mathematical programs. Only by strengthening convexity are we able to guarantee that all local solutions are global, and so approximation iterative steps will in fact produce good approximations to global solutions.

Theorem 5.2 *Suppose that the integrand* $F(t, \mathbf{u}, \mathbf{p})$ *is as in the preceding theorem, and enjoys the joint convexity condition: for every* $t \in [0, T]$, *the function* $(\mathbf{u}, \mathbf{p}) \mapsto F(t, \mathbf{u}, \mathbf{p})$ *is convex as a function of the* $2N$ *variables* $(\mathbf{u}, \mathbf{p})$. *Then every local solution of the corresponding variational problem, under end-point constraints, is also global.*

Once again, we do not want to get into the intricacies of local optimal solutions. To sum up our discussion and establish a parallelism with mathematical programs, let us say that the joint convexity condition of the integrand F with respect to $(\mathbf{u}, \mathbf{p})$ is the property that we should be concerned about in specific examples, because:

- solutions of optimality conditions are (global) solutions of variational problems;
- approximation of optimal solutions through descent techniques will indeed produce true approximate solutions (close to continuous optimal solutions);
- if convexity is strengthened to strict convexity, then optimal solutions are unique.

5.3 Optimality Conditions Under Smoothness

The tool that we are going to describe now is the central part of the chapter. It plays a similar role with respect to variational problems, as KKT conditions for their mathematical programming counterparts. Since we are now working with continuous optimization problems, optimality conditions come in the form of differential equations.

Our plan for the section is the following. We will first argue why the form of optimality conditions is what it is, and where it comes from. Again, we will avoid dealing with fundamental technical issues that are part of a more profound analysis, but apply those to some specific examples seeking to gain a certain familiarity with those optimality conditions. In this context, we talk about the Euler–Lagrange equations of the problem.

Theorem 5.3 *Let the integrand $F(t, \mathbf{u}, \mathbf{p})$ be smooth, and consider the variational problem*

$$\text{Minimize in } \mathbf{u}: \quad I(\mathbf{u}) = \int_0^T F(t, \mathbf{u}(t), \mathbf{u}'(t)) \, dt$$

under the end-point conditions $\mathbf{u}(0) = \mathbf{u}_0$, $\mathbf{u}(T) = \mathbf{u}_T$, for given vectors $\mathbf{u}_0$, $\mathbf{u}_T \in \mathbb{R}^N$. Under further technical assumptions (that we omit but hold for most of the regular problems one encounters in practice), if $\overline{\mathbf{u}}$ is an optimal solution (even a local one) then

$$-[F_{\mathbf{p}}(t, \overline{\mathbf{u}}(t), \overline{\mathbf{u}}'(t))]' + F_{\mathbf{u}}(t, \overline{\mathbf{u}}(t), \overline{\mathbf{u}}'(t)) = \mathbf{0} \text{ in } (0, T), \qquad (5.3)$$

in addition to being feasible $\overline{\mathbf{u}}(0) = \mathbf{u}_0$, $\overline{\mathbf{u}}(T) = \mathbf{u}_T$.

A couple of initial observations are worth stating:

- no partial derivative with respect to t occurs in (5.3);
- (5.3) is in fact a system of N ordinary differential equations in the N-unknowns $\mathbf{u} = (u_1, u_2, \ldots, u_N)$.

Though we do not pretend to write a formal and fully rigorous proof of this important result, it will probably be interesting for readers to have some information about where the Euler–Lagrange equations come from. How do they arise? Assume that $\overline{\mathbf{u}}(t)$ is indeed a true minimizer of our problem, and write down a perturbation, a "variation", of $\overline{\mathbf{u}}$ in the form $\overline{\mathbf{u}}(t) + \varepsilon\mathbf{u}(t)$ where we take $\mathbf{u}$ arbitrarily as long as $\mathbf{u}(0) = \mathbf{u}(T) = \mathbf{0}$ to ensure that such a variation is feasible for every value of ε. For every such ε, this variation is feasible, and since $\overline{\mathbf{u}}(t)$ is the minimizer, we should have

$$I(\overline{\mathbf{u}}) \leq I(\overline{\mathbf{u}} + \varepsilon\mathbf{u})$$

for every value of ε. This means that if we consider the function of a single real variable

$$g(\varepsilon) = I(\overline{\mathbf{u}} + \varepsilon\mathbf{u}),$$

such function has a global minimum at $\varepsilon = 0$. Hence $g'(0) = 0$ (and $g''(0) \geq 0$ as well). How do we compute the derivative $g'(\varepsilon)$? By looking at the formula

$$g(\varepsilon) = \int_0^T F(t, \overline{\mathbf{u}}(t) + \varepsilon\mathbf{u}(t), \overline{\mathbf{u}}'(t) + \varepsilon\mathbf{u}'(t))\, dt,$$

differentiation with respect to ε would yield

$$g'(\varepsilon) = \int_0^T [F_{\mathbf{u}}(t, \overline{\mathbf{u}}(t) + \varepsilon\mathbf{u}(t), \overline{\mathbf{u}}'(t) + \varepsilon\mathbf{u}'(t))\mathbf{u}(t)$$
$$+ F_{\mathbf{p}}(t, \overline{\mathbf{u}}(t) + \varepsilon\mathbf{u}(t), \overline{\mathbf{u}}'(t) + \varepsilon\mathbf{u}'(t))\mathbf{u}'(t)]\, dt.$$

This formula would require a serious discussion about differentiation under the integral sign. We take it for granted, and proceed to conclude that

$$g'(0) = \int_0^T [F_{\mathbf{u}}(t, \overline{\mathbf{u}}(t), \overline{\mathbf{u}}'(t))\mathbf{u}(t) + F_{\mathbf{p}}(t, \overline{\mathbf{u}}(t), \overline{\mathbf{u}}'(t))\mathbf{u}'(t)]\, dt = 0.$$

But the variation $\mathbf{u}(t)$ can be taken in an arbitrary way as long as we ensure that $\mathbf{u}(0) = \mathbf{u}(T) = \mathbf{0}$. For all such $\mathbf{u}$'s, we ought to have

$$\int_0^T [F_{\mathbf{u}}(t, \overline{\mathbf{u}}(t), \overline{\mathbf{u}}'(t))\mathbf{u}(t) + F_{\mathbf{p}}(t, \overline{\mathbf{u}}(t), \overline{\mathbf{u}}'(t))\mathbf{u}'(t)]\, dt = 0.$$

A typical integration by parts in the second term to charge the derivative on the other factor, leads to

$$\int_0^T [F_{\mathbf{u}}(t, \overline{\mathbf{u}}(t), \overline{\mathbf{u}}'(t))\mathbf{u}(t) - F_{\mathbf{p}}(t, \overline{\mathbf{u}}(t), \overline{\mathbf{u}}'(t))'\mathbf{u}(t)]\, dt = 0.$$

Note that the end-point contributions drop out precisely because of the vanishing end-point conditions for the variation $\mathbf{u}(t)$. Thus, we come to the conclusion that, if $\overline{\mathbf{u}}$ is a minimizer of our problem, then

$$\int_0^T [F_{\mathbf{u}}(t, \overline{\mathbf{u}}(t), \overline{\mathbf{u}}'(t)) - F_{\mathbf{p}}(t, \overline{\mathbf{u}}(t), \overline{\mathbf{u}}'(t))']\mathbf{u}(t)\,dt = 0,$$

for every such variation $\mathbf{u}(t)$. Due to this arbitrariness on $\mathbf{u}(t)$, the only way that all of those integrals may vanish is that the factor

$$F_{\mathbf{u}}(t, \overline{\mathbf{u}}(t), \overline{\mathbf{u}}'(t)) - F_{\mathbf{p}}(t, \overline{\mathbf{u}}(t), \overline{\mathbf{u}}'(t))'$$

identically vanish. This is the Euler–Lagrange set of equations.

In order to better understand the nature, and the way to manipulate those Euler–Lagrange equations, we are going to proceed by looking at specific examples of increasing difficulty.

1. The simplest, non-trivial situation corresponds to an integrand F which only depends on the derivative $\mathbf{p}$ so that $F = F(\mathbf{p})$. The functional to be minimized is then

$$I(\mathbf{u}) = \int_0^T F(\mathbf{u}'(t))\,dt$$

   under end-point conditions $\mathbf{u}(0) = \mathbf{u}_0$, $\mathbf{u}(T) = \mathbf{u}_T$. In this case, the Euler–Lagrange equation in (5.3) becomes $\nabla F(\mathbf{u}'(t)) =$ constant independent of t. The affine function passing through the two vectors given at the end-point conditions will always be a solution. It will be the unique solution of the problem, if F is a strictly convex function of $\mathbf{p}$.

2. We next consider the case $F(t, \mathbf{p})$, so that the problem is

$$\text{Minimize in } \mathbf{u}: \quad I(\mathbf{u}) = \int_0^T F(t, \mathbf{u}'(t))\,dt$$

   subject to $\mathbf{u}(0) = \mathbf{u}_0$, $\mathbf{u}(T) = \mathbf{u}_T$. Because the integrand does not depend explicitly upon $\mathbf{u}$, the Euler–Lagrange equation will still be

$$F_{\mathbf{p}}(t, \mathbf{u}'(t)) = \nabla_{\mathbf{p}} F(t, \mathbf{u}'(t)) = \text{constant},$$

   but since we are assuming an explicit dependence of F on t, this time we cannot conclude, in general, that the solution is the affine function through the two end-point vectors. A full analysis is required for each particular situation.

   Possibly, the best known example in this category is the brachistochrone, for which we take (after a suitable identification of variables)

$$F(t, p) = \frac{\sqrt{1 + p^2}}{\sqrt{t}}.$$

The functional to be minimized is

$$I(u) = \int_0^T F(t, u'(t)) \, dt$$

under $u(0) = 0$, $u(T) = u_T$. As indicated above, the Euler–Lagrange equation reduces to

$$\frac{u'(t)}{\sqrt{t(1 + u'(t)^2)}} = \frac{1}{c} \tag{5.4}$$

for c, a constant. After a bit of algebra, we have

$$u'(t)^2 = \frac{t}{c^2 - t}, \quad u(t) = \int_0^t \sqrt{\frac{s}{c^2 - s}} \, ds.$$

The constant c would be selected by demanding

$$u_T = \int_0^T \sqrt{\frac{s}{c^2 - s}} \, ds.$$

This form of the solution is not particularly appealing. One can find a better equivalent form by making an attempt to describe the solution in parametric form with the goal of introducing a good change of variable in the integral defining $u(t)$, and calculating it in a more explicit form. Indeed, if we take

$$s(r) = c^2 \sin^2(r/2) = \frac{c^2}{2}(1 - \cos r),$$

and perform the corresponding change of variables in the definition of $u(t)$, we find

$$u(\tau) = c^2 \int_0^\tau \sin^2(\frac{r}{2}) \, dr = \frac{c^2}{2}(\tau - \sin \tau),$$

where the upper limit in the integral is

$$t = \frac{c^2}{2}(1 - \cos \tau).$$

We, then, see that the parametric form of the optimal profile is

$$(t(\tau), u(\tau)) = C(1 - \cos \tau, \tau - \sin \tau), \quad \tau \in [0, 1],$$

where the constant C is chosen to ensure that $t(1) = T$, $u(1) = u_T$. This family of curves is well-known in Differential Geometry: they are arcs of cycloids, and enjoy quite remarkable properties.

One of the most surprising ones is its tautochrone condition: the transit time of the mass falling along an arc of a cycloid is independent of the point where the mass starts to fall from !! No matter how high or how low you put the mass, they both will employ the same time interval in getting to the lowest point. This is not hard to argue, once we have gone through the above computations. Notice that the cycloid is parameterized by $(t(\tau), u(\tau)) = C(1 - \cos \tau, \tau - \sin \tau)$ for some constant C and $\tau \in [0, 1]$, so that it is easy to realize that $u'(\tau) = t(\tau)$. If we go back to (5.4), and bear in mind this identity, it is easy to find that indeed

$$\sqrt{\frac{t}{1 + u'(t)^2}} = \tilde{c}, \qquad (5.5)$$

a constant. Because in the variable τ the interval of integration for the transit time is constant $[0, 1]$, we see that for the optimal profile, the cycloid, the transit time, whose integrand is the inverse of (5.5), is constant independent of the height u_T. This implies our claim. We will see some numerical approximations, and pictures in the next chapter.

3. The final situation that we would like to discuss in an explicit fashion is that of an integrand $F = F(\mathbf{u}, \mathbf{p})$ depending only on $\mathbf{u}$ and $\mathbf{p} = \mathbf{u}'$. In this case, there is an interesting computation showing that if $\mathbf{u}$ is a solution of (5.3), then

$$F(\mathbf{u}(t), \mathbf{u}'(t)) - \mathbf{u}'(t) \cdot F_{\mathbf{p}}(\mathbf{u}(t), \mathbf{u}'(t)) = \text{constant}. \qquad (5.6)$$

This is elementary. For the scalar case where $N = 1$, this condition is equivalent to (5.3).

One of the most interesting situations of this kind is that of finding the minimal surface of revolution, in which we seek the profile $u(t)$, $0 \le t \le T$, so that the graph of u generates a surface of revolution around the X-axis of minimal area. In analytical terms, we seek to

$$\text{Minimize in } u(t): \quad \int_0^T u(t)\sqrt{1 + u'(t)^2}\, dt$$

under the usual end-point conditions $u(0) = u_0$, $u(T) = u_T$, with $u_0, u_T > 0$. Except for the positive multiplicative constant 2π, the integrand for this problem is coming from elementary Calculus courses. The integrand is then $F(u, p) = u\sqrt{1 + p^2}$. To find the optimal profile, we put, according to (5.6),

$$u(t)\sqrt{1 + u'(t)^2} - u'(t)\frac{u(t)u'(t)}{\sqrt{1 + u'(t)^2}} = c,$$

a constant. After some careful manipulations we arrive at

$$u'(t) = \frac{1}{c}\sqrt{u(t)^2 - c^2}.$$

To perform the integration, we separate variables to have

$$\frac{du}{\sqrt{u^2 - c^2}} = \frac{dt}{c},$$

and a standard hyperbolic trigonometric change of variables leads immediately to the form of the optimal profile

$$u(t) = c \cosh \left(\frac{t}{c} + d \right)$$

for constants c, and d, that are determined to adjust the end-point conditions. We will also see some pictures, through suitable numerical approximations, in the next chapter.

There is a quite interesting discussion about this problem of a clear geometric flavor, involving the relative sizes of T, and u_0, u_T. For some specific regime, the optimal profile just found breaks down, and no optimal profile, with a minimum surface area of revolution, can be found ([20]). Notice that the integrand $F(u, p) = u\sqrt{1 + p^2}$ is not jointly convex in pairs (u, p).

It is worth stating that variational principles enjoy a rich tradition in Mechanics. The paradigmatic situation is that of a functional in the form of an action integral

$$E(\mathbf{u}) = \int_0^T \left[\frac{1}{2} |\mathbf{u}'(t)|^2 - P(\mathbf{u}(t)) \right] dt$$

where the first term accounts for kinetic energy, while P is a potential. This is a more advanced topic (see the classical book [15]).

5.4 Constraints

In this final section, we would like to treat the case in which constraints of different nature, in addition to end-point conditions, are to be respected. In particular, we are interested in understanding how to deal with pointwise conditions.

A first situation of interest, before turning to having more constraints, is to have less constraints. Suppose for instance that in a typical variational problem, just one end-point condition is to be enforced

$$\text{Minimize in } \mathbf{u} : \quad I(\mathbf{u}) = \int_0^T F(t, \mathbf{u}(t), \mathbf{u}'(t)) \, dt$$

under $\mathbf{u}(0) = \mathbf{u}_0$, but there is no restriction about the value of $\mathbf{u}$ at $t = T$. In this case, we say that the optimal solution will seek the natural boundary condition at the right-end point.

Theorem 5.4 *Let the integrand $F(t, \mathbf{u}, \mathbf{p})$ be smooth, and consider the variational problem*

$$\text{Minimize in } \mathbf{u}: \quad I(\mathbf{u}) = \int_0^T F(t, \mathbf{u}(t), \mathbf{u}'(t)) \, dt$$

under the end-point conditions $\mathbf{u}(0) = \mathbf{u}_0$. If $\overline{\mathbf{u}}$ is an optimal solution (even a local one) then

$$-[F_{\mathbf{p}}(t, \overline{\mathbf{u}}(t), \overline{\mathbf{u}}'(t))]' + F_{\mathbf{u}}(t, \overline{\mathbf{u}}(t), \overline{\mathbf{u}}'(t)) = \mathbf{0} \text{ in } (0, T), \qquad (5.7)$$

in addition to

$$\overline{\mathbf{u}}(0) = \mathbf{u}_0, \quad F_{\mathbf{p}}(T, \overline{\mathbf{u}}(T), \overline{\mathbf{u}}'(T)) = \mathbf{0}.$$

The condition

$$F_{\mathbf{p}}(T, \overline{\mathbf{u}}(T), \overline{\mathbf{u}}'(T)) = \mathbf{0}$$

is called the transversality condition or the natural boundary condition at $t = T$. Notice that it involves exactly the field within the first-derivative in the Euler–Lagrange system. It is an interesting exercise to argue the validity of the transversality condition in the context of the integration by parts performed earlier when we discussed where the Euler–Lagrange system of equations came from.

If the restriction at the left-end point is also liberated then we would have both natural boundary conditions

$$F_{\mathbf{p}}(0, \overline{\mathbf{u}}(0), \overline{\mathbf{u}}'(0)) = \mathbf{0}, \quad F_{\mathbf{p}}(T, \overline{\mathbf{u}}(T), \overline{\mathbf{u}}'(T)) = \mathbf{0}.$$

Because feasible paths $\mathbf{u}$ have various components, it could happen in problems that some components require both end-point conditions, while some others may require just one (or none) end-point condition. In that case, the corresponding natural boundary condition would be applied just to the appropriate components.

There is a different kind of constraints that some times needs to be enforced so that problems may have a clear meaning. These are global constraints expressed in terms of further integrals of the form

$$\int_0^T \mathbf{G}(t, \mathbf{u}(t), \mathbf{u}'(t)) \, dt = \mathbf{g}, \qquad (5.8)$$

for a vector mapping $\mathbf{G}(t, \mathbf{u}, \mathbf{p}) : [0, T] \times \mathbb{R}^N \times \mathbb{R}^N \to \mathbb{R}^m$, and $\mathbf{g} \in \mathbb{R}^m$. Constraints could also be expressed in the form of inequalities, though we will restrict attention in this chapter to integral, global constraints in the form of equalities. After our experience in Chap. 3, it will not be surprising that a vector of multipliers $\mathbf{z} \in \mathbb{R}^m$ needs to be introduced, and write the Euler–Lagrange equation (optimality conditions) for the modified integrand

$$\tilde{F}(t, \mathbf{u}, \mathbf{p}) = F(t, \mathbf{u}, \mathbf{p}) + \mathbf{z} \cdot \mathbf{G}(t, \mathbf{u}, \mathbf{p}).$$

Namely, we should look for the solution of the problem

$$-[\tilde{F}_{\mathbf{p}}(t, \overline{\mathbf{u}}(t), \overline{\mathbf{u}}'(t))]' + \tilde{F}_{\mathbf{u}}(t, \overline{\mathbf{u}}(t), \overline{\mathbf{u}}'(t)) = \mathbf{0} \text{ in } (0, T),$$

subject to the appropriate end-point conditions. In addition, integral constraints (5.8) ought to be imposed to determine the precise value of the vector of multipliers $\mathbf{z}$, and the full optimal solution $\mathbf{u}$.

Example 5.4 The best known example under an integral constraint is that of the hanging cable as described at the beginning of the chapter. Recall that we pretend to

$$\text{Minimize in } u(t): \quad \int_0^H u(t)\sqrt{1 + u'(t)^2}\, dt$$

under the constraints

$$u(0) = u(H) = 0, \quad \int_0^H \sqrt{1 + u'(t)^2}\, dt = L. \tag{5.9}$$

The modified integrand would be

$$\tilde{F}(u, p) = u\sqrt{1 + p^2} + z\sqrt{1 + p^2} = (u + z)\sqrt{1 + p^2}.$$

We actually realize that this integrand is exactly the same that the one for the minimal surfaces of revolution, if we implement the change $U(t) = u(t) + z$. Note that $U'(t) = u'(t)$. Thus, bearing in mind the computations we went through for that problem, we conclude

$$u(t) = c \cosh\left(\frac{t}{c} + d\right) - z$$

where now the three constants c, d, and z are found by requiring that this u complies with the three constraints in (5.9). The important thing, beyond this final adjustments of constants, is that the optimal profile is a hyperbolic cosine curve, which is usually referred to as a catenary. We will numerically calculate some approximations of the optimal solution for this problem.

Finally, there is also the important situation in which we are to respect a certain constraint (or several of them) for every value of the independent variable t. In this case, the multiplier $z(t)$ becomes a function of time too. Constraints could come in the form of equality or in the form of inequalities. The most important example of the first is the state law in an optimal control problem, while there are several interesting situations for the second possibility. We will treat various examples of both categories from an approximating viewpoint.

5.5 Final Remarks

We have hardly started to scratch the surface of the field Calculus of Variations. Our intention has been to better appreciate the relevance of these optimization problems, and timidly state the important concepts at the heart of a full understanding: convexity for existence of solutions, and optimality conditions in the form of the Euler–Lagrange equations. Readers can explore some of the references in the bibliography section.

We will also pay some attention to approximation techniques so that we may count on a affordable technique which, with a bit of effort, may help in finding reasonable approximations of optimal solutions for variational problems. This is one main aim of this book.

5.6 Exercises

5.6.1 Exercises to Support the Main Concepts

1. Look at the variational problem consisting in minimizing the integral

$$E(u) = \frac{1}{2} \int_0^1 u'(x)^2 \, dx$$

   among those functions $u(x)$ such that $u(0) = 0$, $u(1) = 1$.

   (a) It is very easy to check that the family $u_t(x) = x^t$ for each $t > 0$ is feasible for that problem. Find the value of t which is best.
   (b) The family of functions

$$u(x) = x + \sin(n\pi x)$$

   for each integer n is also feasible. Find the best n.
   (c) For an arbitrary $v(x)$, the function

$$u(x) = x + x(1 - x)v(x)$$

   is admissible. Can you find the best $v(x)$?

2. Consider the functional

$$\int_0^1 (u(x)^2 - xu(x) + 3) \, dx.$$

(a) Find the function $u(x)$ minimizing it among all possible functions under no constraint.

(b) Which is the minimizer if we demand $u(0) = 0$, $u(1) = 1$?

(c) Find the minimizer $u(x)$ under those same end-point constraints for the functional

$$\int_0^1 \left[\frac{1}{2} u'(x)^2 + (u(x)^2 - xu(x) + 3) \right] dx.$$

(d) How do calculation change for the functional

$$\int_0^1 \left[\frac{\varepsilon}{2} u'(x)^2 + (u(x)^2 - xu(x) + 3) \right] dx,$$

regarded ε as a small parameter? Draw a draft of the minimizer for decreasing values of ε.

3. Find the optimality equation for a problem depending on second derivatives

$$\int_0^L F(x, u(x), u'(x), u''(x)) dx$$

where $F(x, u, p, q)$ is a smooth function of four variables, under end-point conditions fixing the values of u and u'. Apply it to the particular situation

$$\int_0^{\pi/\sqrt{2}} [u''(x)^2 + u(x)^2 + x^2] dx$$

with

$$u(0) = 1, \quad u'(0) = \sqrt{2}/2, \quad u(\pi/\sqrt{2}) = 0, \quad u'(\pi/\sqrt{2}) = -\frac{\sqrt{2}}{2} e^{\pi/2}.$$

4. Consider the functional

$$\int_0^1 F(x, u(x), u'(x)) dx.$$

Study optimality conditions for the minimizer $u(x)$ among all those periodic functions $u(0) = u(1)$. Apply your conclusions to find the minimizer for the particular

$$F(x, u, p) = \frac{1}{2} p^2 + \frac{1}{2} (u - x)^2.$$

5. For the functional in the previous exercise, write a 1-periodic function $u(x)$ in Fourier series of cosines

$$u(x) = \sum_{j=0}^{\infty} a_j \cos(2\pi jx),$$

and look at the minimization problem

$$I(\{a_j\}) = \int_0^1 F\left(x, \sum_{j=0}^{\infty} a_j \cos(2\pi jx), -2\pi \sum_{j=1}^{\infty} ja_j \sin(2\pi jx)\right) dx.$$

Try to understand optimality conditions with respect to the Fourier coefficients in this form, and check that you reach the same conclusion.

6. For a variational problem where pairs of functions $(u(x), v(x))$ compete for the minimum of

$$\int_0^1 F(x, u(x), v(x), u'(x), v'(x)) \, dx,$$

write the optimality conditions under end-point constraints. Apply those ideas to the quadratic problem

$$\int_0^1 \frac{1}{2}(u'(x), v'(x)) A \begin{pmatrix} u'(x) \\ v'(x) \end{pmatrix} dx$$

where $\mathbf{A}$ is a symmetric, positive definite 2×2-matrix, and we demand

$$(u(i), v(i)) = (u_i, v_i), \quad i = 0, 1.$$

7. Consider the functional

$$\int_0^1 F(x, u(x), u'(x)) \, dx.$$

For each pair $(t, y) \in (0, 1) \times \mathbb{R}$, define the so-called value function of the problem

$$v(t, y) = \min_u \{\int_t^1 F(x, u(x), u'(x)) \, dx : u(t) = y, u(1) = 0\}.$$

Assuming that this function is as smooth as calculations may require to be, conclude that (Hamilton–Jacobi–Bellman equation)

$$-\frac{\partial v}{\partial t}(t, y) = \min_z \{F(x, u, z) + z\frac{\partial v}{\partial x}(t, x)\}.$$

8. There is a rich tradition in variational methods in Mechanics. If $\mathbf{q}(t)$ represents the state of a certain mechanical system with velocities $\mathbf{q}'(t)$, then the dynamics

of the system evolves in such a way that the Euler–Lagrange equations for the action functional

$$\int_0^T L(\mathbf{q}(t), \mathbf{q}'(t)) \, dt$$

hold. The integrand $L(\mathbf{q}, \mathbf{z}) : \mathbb{R}^N \times \mathbb{R}^N \to \mathbb{R}$ is called the Lagrangian of the system. The Hamiltonian $H(\mathbf{q}, \mathbf{p})$ is defined through the formula

$$H(\mathbf{q}, \mathbf{p}) = \sup_{\mathbf{z} \in \mathbb{R}^N} \{\mathbf{z} \cdot \mathbf{p} - L(\mathbf{q}, \mathbf{z})\}.$$

Argue, formally, that the system will evolve according to the hamiltonian equations

$$\mathbf{p}'(t) = -\frac{\partial H}{\partial \mathbf{q}}(\mathbf{q}(t), \mathbf{p}(t)), \quad \mathbf{q}'(t) = \frac{\partial H}{\partial \mathbf{p}}(\mathbf{q}(t), \mathbf{p}(t)),$$

and that H is preserved through integral trajectories.

9. Show that the variational problem

$$\text{Minimize in } u(t) : \quad \int_0^1 [tu'(t) + u(t)^2] \, dt$$

under $u(0) = 0$, $u(1) = 1$, cannot have a differentiable minimizer.

10. A functional

$$\int_a^b F(t, u(t), u'(t)) \, dt$$

whose value actually depends only on the values of u at the end-points $t = a$, $t = b$ is called a null-Lagrangian. Check that for an arbitrary, smooth, real function f, the functional

$$\int_a^b u'(t)[f(u(t)) + u(x)f'(u(t))] \, dt$$

is a null-Lagrangian. More generally, for a smooth arbitrary function of two variables $F(u, t)$, the functional

$$\int_a^b \left[\frac{\partial F}{\partial u}(u(t), t)u'(t) + \frac{\partial F}{\partial t}(u(t), t) \right] dt$$

is a null-Lagrangian. What happens with the corresponding E–L equations?

11. Variational problems in higher dimension are much harder, especially vector problems.

(a) Consider the Dirichlet functional

$$I(u) = \frac{1}{2} \int_\Omega |\nabla u(\mathbf{x})|^2 \, d\mathbf{x}, \quad \Omega \subset \mathbb{R}^2.$$

Write the E–L equation for it, by using the techniques of "making variations" as in the one-dimensional situations, under a Dirichlet boundary condition $u = u_0$ for a certain fixed boundary datum $u_0(\mathbf{x})$.

(b) Let now

$$\mathbf{u}(\mathbf{x}) = (u_1(x_1, x_2), u_2(x_1, x_2)) : \Omega \subset \mathbb{R}^2 \to \mathbb{R}^2,$$

and consider the functional

$$\int_\Omega \det \nabla \mathbf{u}(\mathbf{x}) \, d\mathbf{x}, \quad \det \nabla \mathbf{u}(\mathbf{x}) = \frac{\partial u_1}{\partial x_1} \frac{\partial u_2}{\partial x_2} - \frac{\partial u_1}{\partial x_2} \frac{\partial u_2}{\partial x_1}.$$

Write the E–L system, i.e. the E–L equation for each component u_i, $i = 1, 2$. What have you found? What interpretation do you suggest? Can you convince yourself otherwise of this interpretation?

5.6.2 Practice Exercises

1. Write the E–L equation for the following functionals

$$\int_a^b [y'(x)^2 + e^{y(x)}] \, dx, \quad \int_a^b u'(x)u(x) \, dx, \quad \int_0^1 t^2 y'(t)^2 \, dt.$$

2. Find the optimal solution of the problem

$$\text{Minimize in } y(x) : \quad \int_0^1 (y'(x)^2 + 2y(x)^2) e^x \, dx$$

under $y(0) = 0$, $y(1) = e - e^{-1}$.

3. Look for the minimizer of the functional

$$\int_0^1 [u'(x)^2 + u(x)^2 + u(x)u'(x)] \, dx$$

subject to the end-point conditions $u(0) = 0$, $u(1) = 1$.

4. Calculate the minimal possible value of the integrals

$$\int_0^{\log 2} [(y'(x) - 1)^2 + y(x)^2] \, dx.$$

5. Look at the function

$$f(s) = \min \left\{ \int_0^{\log 2} (y'(x)^2 + y'(x) + y(x)^2) \, dx, \ y(0) = 0, \ y(\log 2) = s \right\}.$$

Find an explicit expression for $f(s)$, and determine the optimal solution of the problem

$$\min \left\{ \int_0^{\log 2} (y'(x)^2 + y'(x) + y(x)^2) \, dx, \ y(0) = 0 \right\}.$$

Compare this result with the solution obtained directly through the transversality or natural boundary condition.

6. Find the function that minimizes the integral

$$J(y) = \int_0^1 y'(x)^2 \, dx, \ y(0) = 0, \ y(1) = 0,$$

under the integral constraint

$$\int_0^1 y(x) \, dx = 1.$$

Seek the same answer by putting $y(t) = z'(t)$, with $z(0) = 0$, $z(1) = 1$.

7. In a transit problem, like the one for the brachistochrone, velocity is assumed to be proportional to the distance to ground, not to the square root of that distance (as in the brachistochrone). Figure out what kind of a curve is the optimal profile this time. You can set end-point conditions $u(0) = 0$, $u(1) = 1$.

8. Consider the functional

$$\int_0^1 [u'(x)^2 + u(x)u'(x)]^2 \, dx$$

under the only constraint $u(1) = 0$.

(a) Give a specific feasible function $u_1(x)$ and compute the value of the functional for it.

(b) Give a second feasible function $u_2(x)$, compute its cost, and compare it to the previous one.

(c) Can you ensure that the solution of the Euler–Lagrange equation will provide the optimal function? Why or why not?

(d) How many minimizers are there? Find all of them.

9. Argue why the variational problem

$$\text{Minimize in } y(x) : \quad \int_0^1 [xy'(x) + y'(x)^2] \, dx$$

under $y(0) = 1$ has a unique solution. Find it.

10. We are given the following functional

$$\int_0^1 (x'(t) - x(t) - 1)^2 \, dt$$

under the condition $x(0) = 1$. Write the associated Euler–Lagrange equation, the transversality condition for $t = 1$, and argue that the only minimizer is the unique solution of the Cauchy problem

$$x'(t) = x(t) + 1 \text{ in } (0, 1), \quad x(0) = 1.$$

Find such function.

11. Consider the functional

$$\int_0^1 \left[u'(x)^2 + \frac{1}{u'(x)} \right] dx$$

under the end-point conditions $u(0) = 0$, $u(1) = 1$, $u'(x) > 0$. Check if convexity conditions hold, and if existence of a unique minimizer can be guarantee. Find the minimizer(s).

12. The problem

$$-u'' + g(u) = 0 \text{ in } (0, 1), \quad u(0) = 0, u(1) = 1,$$

corresponds to the Euler–Lagrange equation of some functional.

(a) If $g(u) = u - 1$, can you figure out such functional? Is the solution of that differential problem the minimizer of the functional?
(b) Same questions for $g(u) = u^2$.
(c) Determine the functional for an arbitrary function g.
(d) What properties on g would be required to ensure that the solution of the Euler–Lagrange problem would correspond to the minimizer of the functional.

13. Look at the variational problem

$$\min_{y \in \mathscr{A}} \int_0^1 \left(xy'(x) + y'(x)^2 \right) dx$$

where $\mathscr{A} = \{y \in \mathscr{C}^1([0, 1]) : y(0) = 1\}$ is the set of functions continuously differentiable taking on the value 1 at zero. Find its optimal solution in the following situations:

(a) if $y(1)$ is free;
(b) if $y(1) \geq 1$,
(c) if $y(1) \leq \frac{1}{2}$.

14. Let us consider the two following, apparently unrelated optimization problems

$$\frac{1}{2} \int_0^1 u'(t)^2 \, dt$$

subject to the restrictions

$$0 \le u(0) \le 1, u(1) = 1, \quad \int_0^1 u(t) \, dt \le \frac{1}{2},$$

and the non-linear program

$$\text{Minimize in } (x, y): \quad x^2 + 3xy + 3y^2$$

under the constraints

$$0 \le x + 2y \le 2, \quad 3 \le 2x + 3y.$$

(a) Is the feasible region of this last problem bounded? Is it convex? Is the cost a convex function? Is the infimum value attained? Argue your answers.
(b) Compute the point of minimum of this program.
(c) By making use of the Euler–Lagrange equation, transform the variational problem into the non-linear program.
(d) Through the optimal solution of the latter, find the optimal function of the former.

5.6.3 Case Studies

1. Find the shortest path(s) joining two points in the unit sphere in $\mathbb{R}^3$.
2. For the set of all differentiable curves joining $(0, 0)$ and $(1, 1)$:

   (a) Find the one of a constant length B that encloses the largest area with respect to the horizontal axis.
   (b) Determine the curve of smallest length enclosing a fixed area B.

3. Let us look at the surface $\mathbf{S}$ in $\mathbb{R}^3$ of equation $y = \cosh x$ (z, free). We would like to determine some of its geodesics.

   (a) Write down the functional providing the length of each curve σ contained in $\mathbf{S}$.
   (b) For the two points $(0, 1, 0)$ and $(1, \cosh 1, 1)$, are the curves

   $$\sigma_1(t) = (0, 1, 0) + t(1, \cosh 1 - 1, 1), \quad t \in [0, 1],$$

and
$$\sigma_2(t) = (t, \cosh t, t^2), \quad t \in [0, 1],$$

feasible for the problem? Why?
(c) Find the length of those two curves. Which one has a smaller length?
(d) Find the curve of shortest length between those two points.

4. An elastic string of unit length is supported on its two end-points at the same height $y = 0$. Its equilibrium profile under the action of a vertical load of density $g(x) = -x(1 - x)$ for $x \in [0, 1]$ is the one minimizing internal energy which is approximated, given that the values of g are reasonably small, by the functional

$$E(y) = \int_0^1 [\frac{1}{2}y'(x)^2 + g(x)y(x)]\, dx.$$

(a) Is the profile itself $y(x) = -x(1 - x)$ admissible? Why? What is its internal energy?
(b) Find a feasible profile different from the previous one, and compute its internal energy.
(c) Determine the optimal profile minimizing $E(y)$, and draw a picture of it.

5. An elastic bar of unit length hangs freely from the ceiling. When a weight W is put at the free end, each transversal section at length x from the ceiling undergoes a vertical displacement $u(x)$ downwards. The energy associated with such a configuration $u(x)$ is given by the integral

$$\int_0^1 \left[\frac{(1 + x)}{2} u'(x)^2 - W u'(x) \right] dx$$

where the term $1 + x$ is the elastic coefficient of the section corresponding to x. We postulate that the situation in equilibrium $u(x)$, after putting the weight W, is the one minimizing the previous integral.

(a) What must end-point conditions be at the free end-point and at the one in the ceiling?
(b) Is $u(x) = (W/4)x^2$ feasible? Is $u(x) = (W/2)x$ admissible? Find the energy of these two, if they are feasible.
(c) Find the deformation of minimal energy.

6. Let $(u(t), v(t))$ be the two components of a plane, closed curve σ parameterized in such a way that

$$u(a) = u(b) = 0, \quad v(a) = v(b) = 0.$$

The length of σ is given by

$$L(\sigma) = \int_a^b \sqrt{u'(t)^2 + v'(t)^2} \, dt,$$

while the area enclosed by it is, according to Green's theorem,

$$A(\sigma) = \int_a^b (u(t)v'(t) - u'(t)v(t)) \, dt.$$

We would like to find the optimal curve for the problem

Minimize in σ : $L(\sigma)$ under $A(\sigma) = \alpha, \sigma(a) = \sigma(b) = (0, 0)$.

Argue why the optimal curve needs to be a circle, and deduce the isoperimetric inequality

$$L(\sigma)^2 - 4\pi A(\sigma) \geq 0$$

valid for every closed curve σ in the plane.

7. Optimal profile for Newton's solid of revolution. A solid of revolution generated by the graph of a function $y(x)$ around the Y-axis is moving downwards in a fluid with constant velocity v. End-point conditions are $y(0) = 0$, $y(1) = 1$. The drag coefficient of such a solid (a measure of the resistance the fluid opposes to movement) is given by

$$D(y) = \int_0^1 \frac{2x}{1 + y'(x)^2} \, dx.$$

   (a) Check that D for $y = x$, the line, and $y = 1 - \sqrt{1 - x^2}$, the circle, has the same value.
   (b) What are the convexity properties of the integrand?
   (c) Write the E–L equation for the optimal profile.
   (d) Change the functional to

$$\tilde{D}(y) = \int_0^1 2x \log(1 + y'(x)^2) \, dx,$$

   and check for the convexity of the integrand, and the E–L equation. Go in the solution process as far as you can.

8. Standard harmonic oscillator. The integrand

$$L(\mathbf{x}, \mathbf{x}') = \frac{1}{2} m |\mathbf{x}'|^2 - \frac{1}{2} k |\mathbf{x}|^2$$

is the Lagrangian for a simple harmonic oscillator. The corresponding integral is the action integral.

(a) Find the corresponding hamiltonian $H(\mathbf{p}, \mathbf{q})$.

(b) Write the equations of motion.

(c) Reflect on the convexity properties of the Lagrangian, and conclude that motion takes place along equilibria of the action integral but not along minimizers.

9. The Lagrangian

$$L(\mathbf{x}, \mathbf{x}') = \frac{1}{2}m|\mathbf{x}'|^2 - V(\mathbf{x})$$

corresponds to motion generated by a potential $V(\mathbf{x})$. Write the equations of motion, and argue that motion cannot escape from a valley of the potential, a potential well.

Chapter 6
Numerical Approximation of Variational Problems

6.1 Importance of the Numerical Treatment

We have realized pretty soon that finding optimal solutions for optimization problems, even for moderate-sized cases, is an impossible task by hand, or in a symbolic, analytical manner. This is even more so for variational problems, and optimal control problems. It is therefore necessary to have a mechanism within reach that may be used to find reliable approximations for optimal solutions of problems.

We will follow the same general procedure as with mathematical programming problems: we first focus on mastering the situation without constraints, except possibly for end-point conditions, and then investigate the situation where point wise constraints are to be respected. Exactly the same driving ideas we utilized for mathematical programs are now helpful as well. We first explore steepest descent strategies, and then use these to deal with constraints through iterative procedures.

6.2 Approximation Under Smoothness

We first focus on approximating the optimal solution of a typical variational problem like

$$\text{Minimize in } \mathbf{u}: \quad I(\mathbf{u}) = \int_0^T F(t, \mathbf{u}(t), \mathbf{u}'(t)) \, dt$$

subject to end-point conditions either at both values $t = 0$, $t = T$, or just at the initial value $t = 0$. Our main assumption here is the smoothness of the integrand $F : [0, T] \times \mathbb{R}^N \times \mathbb{R}^N \to \mathbb{R}$ with respect to all of its variables $(t, \mathbf{u}, \mathbf{p})$.

Our proposal is a typical iterative descent mechanism, much in the same way as with mathematical programs. If you recall the discussion in Sect. 4.1, each iterative step amounts to finding a descent direction, and deciding on the step size. Exactly in the same way, we proceed to treat these two ingredients for a variational problem. As

© Springer International Publishing AG 2017
P. Pedregal, *Optimization and Approximation*, UNITEXT 108,
DOI 10.1007/978-3-319-64843-9_6

a matter of fact, once we have the descent direction vector $\mathbf{U}$, treated in the subsection that follows, the discussion on how to select the step size is exactly the same as that in Sect. 4.1.2. Note that the decision on the step size is based on evaluating the cost functional I on vectors of the form $\mathbf{u} + \tau \mathbf{U}$, for various values of τ, and the directional derivative of $\langle I'(\mathbf{u}), \mathbf{U} \rangle$. The only precaution is that the evaluation of the integral functional I has to be approximated through standard quadrature rules of reasonable accuracy.

We will therefore focus our efforts on finding a good way to approximate steepest descent directions (vector functions) $\mathbf{U}$ for I, and refer to Sect. 4.1.2 for a similar discussion about how to choose the step size.

6.2.1 The Descent Direction

Suppose we have a feasible vector function $\mathbf{u}$, complying with end-point conditions $\mathbf{u}(0) = \mathbf{u}_0, \mathbf{u}(T) = \mathbf{u}_T$. For definiteness, we will focus first on such a situation. Let $\mathbf{U}$ be a perturbation of $\mathbf{u}$, just as in the derivation of the Euler-Lagrange equations explained earlier, which will respect the conditions $\mathbf{U}(0) = \mathbf{U}(T) = \mathbf{0}$ so that the perturbed path $\mathbf{u}(t) + \varepsilon \mathbf{U}(t)$ will comply with end-point conditions for every real ε. We compute the derivative of the real function

$$\phi(\varepsilon) = \int_0^T F(t, \mathbf{u}(t) + \varepsilon \mathbf{U}(t), \mathbf{u}'(t) + \varepsilon \mathbf{U}'(t)) \, dt$$

and evaluate it at $\varepsilon = 0$, to find the directional derivative of the functional I at $\mathbf{u}$ in the direction of $\mathbf{U}$. Namely,

$$\langle I'(\mathbf{u}), \mathbf{U} \rangle = \phi'(0).$$

Under smoothness hypotheses on F, this last derivative becomes

$$\langle I'(\mathbf{u}), \mathbf{U} \rangle = \int_0^T \left[F_{\mathbf{u}}(t, \mathbf{u}(t), \mathbf{u}'(t)) \cdot \mathbf{U}(t) + F_{\mathbf{p}}(t, \mathbf{u}(t), \mathbf{u}'(t)) \cdot \mathbf{U}'(t) \right] dt. \quad (6.1)$$

If we insist in finding the steepest descent direction $\mathbf{U}$ with respect to the quadratic norm

$$\|\mathbf{U}\|^2 = \int_0^T |\mathbf{U}'(t)|^2 \, dt, \quad \mathbf{U}(0) = \mathbf{U}(T) = \mathbf{0},$$

then we would have to find such a direction by solving, in turn, the variational problem of minimizing in $\mathbf{U}$ the functional

$$\int_0^T \left[\frac{1}{2} |\mathbf{U}'(t)|^2 + F_{\mathbf{u}}(t, \mathbf{u}(t), \mathbf{u}'(t)) \cdot \mathbf{U}(t) + F_{\mathbf{p}}(t, \mathbf{u}(t), \mathbf{u}'(t)) \cdot \mathbf{U}'(t) \right] dt$$

subject to the appropriate end-point conditions $\mathbf{U}(0) = \mathbf{U}(T) = \mathbf{0}$. Let us stress the fact that this is a regular, quadratic variational problem in the path $\mathbf{U}$. The underlying curve $\mathbf{u}$ is given, and fixed here. The optimal $\mathbf{U}$ will be the solution of the corresponding Euler-Lagrange equations for $\mathbf{U}$

$$- [\mathbf{U}'(t) + F_{\mathbf{p}}(t, \mathbf{u}(t), \mathbf{u}'(t))]' + F_{\mathbf{u}}(t, \mathbf{u}(t), \mathbf{u}'(t)) = 0 \text{ in } (0, T) \tag{6.2}$$

together with $\mathbf{U}(0) = \mathbf{U}(T) = \mathbf{0}$. The form of this system in $\mathbf{U}$ is so special, that its unique solution can be given in an explicit, closed form.

Lemma 6.1 *For the functional I above with smooth integrand $F(t, \mathbf{u}, \mathbf{p})$, the steepest descent direction (with respect to end-point conditions) of I at $\mathbf{u}$ is given by*

$$\mathbf{U}(t) = \frac{t - T}{T} \int_0^T \left[\overline{F}_{\mathbf{p}}(s) - (T - s)\overline{F}_{\mathbf{u}}(s) \right] ds$$

$$+ (t - T) \int_0^t \overline{F}_{\mathbf{u}}(s) \, ds$$

$$+ \int_t^T \left[\overline{F}_{\mathbf{p}}(s) - (T - s)\overline{F}_{\mathbf{u}}(s) \right] ds,$$

where

$$\overline{F}_{\mathbf{u}}(s) = F_{\mathbf{u}}(s, \mathbf{u}(s), \mathbf{u}'(s)), \quad \overline{F}_{\mathbf{p}}(s) = F_{\mathbf{p}}(s, \mathbf{u}(s), \mathbf{u}'(s)).$$

The proof amounts to checking carefully that this path $\mathbf{U}(t)$ complies with all of the conditions in (6.2). This is a very good exercise. The formula given in this statement furnishes the (steepest) descent direction at any given $\mathbf{u}$, under vanishing end-point conditions at both end points. In practice, the integrals occurring in the formula need to be approximated by typical quadrature rules.

It is also interesting to check the local decay of the functional I when we move from $\mathbf{u}$ in the steepest descent direction $\mathbf{U}$ just computed. This amounts to computing (6.1) when $\mathbf{U}$ is precisely the solution of (6.2). After multiplying (6.2) by $\mathbf{U}$ itself, and performing an elementary integration by parts (note that end-point contributions drop out), we find that

$$\langle I'(\mathbf{u}), \mathbf{U} \rangle = - \int_0^T |\mathbf{U}'(t)|^2 \, dt.$$

This quantity will remain strictly negative until the descent direction $\mathbf{U}$ vanishes, i.e. until $\mathbf{u}$ is a solution of the Euler-Lagrange system.

If the starting time T_0 is not zero, then a slight modification of the formula in the last lemma provides the steepest descent direction, namely,

$$\mathbf{U}(t) = \frac{t - T}{T - T_0} \int_{T_0}^{T} \left[\overline{F}_\mathbf{p}(s) - (T - s)\overline{F}_\mathbf{u}(s) \right] ds$$

$$+ (t - T) \int_{T_0}^{t} \overline{F}_\mathbf{u}(s) \, ds$$

$$+ \int_{t}^{T} \left[\overline{F}_\mathbf{p}(s) - (T - s)\overline{F}_\mathbf{u}(s) \right] ds,$$

where $F_\mathbf{u}(s)$ and $F_\mathbf{p}(s)$ are as before.

If our variational problem only demands for the left-end point condition $\mathbf{u}(0) = \mathbf{u}_0$, but $\mathbf{u}(T)$ is free, then we must supply the natural boundary condition at the right end-point for the variational problem defining $\mathbf{U}$

$$\mathbf{U}'(T) + F_\mathbf{p}(T, \mathbf{u}(T), \mathbf{u}'(T)) = 0.$$

The problem determining the descent direction would be this time

$$-[\mathbf{U}'(t) + F_\mathbf{p}(t, \mathbf{u}(t), \mathbf{u}'(t))]' + F_\mathbf{u}(t, \mathbf{u}(t), \mathbf{u}'(t)) = 0 \text{ in } (0, T),$$
$$\mathbf{U}(0) = 0, \quad \mathbf{U}'(T) + F_\mathbf{p}(T, \mathbf{u}(T), \mathbf{u}'(T)) = 0.$$

There is a similar form of the unique solution of this problem in closed form, which is even simpler than the previous one.

Lemma 6.2 *For the functional I above with smooth integrand $F(t, \mathbf{u}, \mathbf{p})$, the steepest descent direction of I at $\mathbf{u}$, under a left-end point condition, is given by*

$$\mathbf{U}(t) = -\int_{0}^{t} [\overline{F}_\mathbf{p}(s) + s\overline{F}_\mathbf{u}(s)] ds - t \int_{t}^{T} \overline{F}_\mathbf{u}(s) \, ds.$$

where, as before,

$$\overline{F}_\mathbf{u}(s) = F_\mathbf{u}(s, \mathbf{u}(s), \mathbf{u}'(s)), \quad \overline{F}_\mathbf{p}(s) = F_\mathbf{p}(s, \mathbf{u}(s), \mathbf{u}'(s)).$$

It is also straightforward to check that this $\mathbf{U}$ is indeed the solution of the problem for the steepest descent direction. Similar observations as above apply to check the local decay of the functional, and the variant of the descent direction $\mathbf{U}$ when the starting time if some T_0 different from zero.

One can then have all possible variations: from having a natural boundary condition only at the left-end point, to having a natural boundary condition at both end-points, passing through the possibility of having different end-point conditions for different components of $\mathbf{u}$. We leave to readers finding the appropriate formulae for those new situations that problems may require.

6.3 Approximation Under Constraints

We focus in this section on how to efficiently approximate the optimal solution of a variational problem under pointwise constraints.

The model problem is

$$\text{Minimize in } \mathbf{u} : \quad \int_0^T F(t, \mathbf{u}(t), \mathbf{u}'(t)) \, dt$$

under

$$\mathbf{u}(0) = \mathbf{u}_0, \mathbf{u}(T) = \mathbf{u}_T, \quad \mathbf{G}(\mathbf{u}(t), \mathbf{u}'(t)) \leq 0 \text{ for all } t \in [0, T].$$

Here $\mathbf{u}_0, \mathbf{u}_T$ are given vectors in $\mathbb{R}^N$, while $F(t, \mathbf{u}, \mathbf{p}) : [0, T] \times \mathbb{R}^N \times \mathbb{R}^N \to \mathbb{R}$ is the integrand for the cost functional, and the smooth mapping $\mathbf{G}(\mathbf{u}, \mathbf{p}) : \mathbb{R}^N \times \mathbb{R}^N \to \mathbb{R}^m$ is used to set up constraints.

We will pursue a strategy similar to the one we used for mathematical programming problems in Sect. 4.2 with the hope that we will find the same encouraging results. It is indeed so. The justification of why that same perspective yields also efficient results for variational problems is more involved, though not essentially more complicated, and we will not insist on it here, but trust that our good results are a good indication that it should be so.

We therefore introduce auxiliary vectors $\mathbf{v}(t) : [0, T] \to \in \mathbb{R}^m$, as many as constraints we are to respect, playing the role of multipliers, and concentrate on the master functional

$$J(\mathbf{u}, \mathbf{v}) = \int_0^T \left[F(t, \mathbf{u}(t), \mathbf{u}'(t)) + \sum_{k=1}^m e^{v^{(k)}(t) G_k(\mathbf{u}(t), \mathbf{u}'(t))} \right] dt.$$

We are setting here $\mathbf{v} = (v^{(k)})$, and $\mathbf{G} = (G_k)$ for the various components.

Our basic algorithm proceeds in a two-step iterative process, the first of which consists in finding the optimal solution for an unconstrained (except for conditions at end-points when applicable) variational problem, and the second is an update rule for multipliers. More specifically:

1. Initialization. Take $\mathbf{u}_0(t)$ any path complying with end-point conditions, but otherwise arbitrary, and $\mathbf{v}(t)$ with arbitrary, strictly positive components, for instance, $\mathbf{v} \equiv \mathbf{1}$.
2. Iterative step in two parts. Let $(\mathbf{u}_j, \mathbf{v}_j)$ be the j-th pair.

   (a) Approximate the optimal solution of the unconstrained variational problem

   $$\text{Minimize in } \mathbf{u} : \quad J(\mathbf{u}, \mathbf{v}_j)$$

under the appropriate end-point conditions. Recall that

$$J(\mathbf{u}, \mathbf{v}_j) = \int_0^T \left[F(t, \mathbf{u}(t), \mathbf{u}'(t)) + \sum_{k=1}^m e^{v_j^{(k)}(t) G_k(\mathbf{u}(t), \mathbf{u}'(t))} \right] dt. \quad (6.3)$$

Let $\mathbf{u}_{j+1}$ be such approximation, found as a result of applying the steepest descent strategy described earlier, starting from $\mathbf{u}_j$.

(b) Stopping and update. Check if all of the products

$$v_j^{(k)}(t) G_k(\mathbf{u}_{j+1}(t), \mathbf{u}'_{j+1}(t))$$

are sufficiently small all over the interval $[0, T]$. If they are, stop, and take $\mathbf{u}_{j+1}$ as a good candidate for the solution of the constrained variational problem. If some products are not yet sufficiently small somewhere in the interval $[0, T]$, update

$$v_{j+1}^{(k)}(t) = e^{v_j^{(k)}(t) G_k(\mathbf{u}_{j+1}(t), \mathbf{u}'_{j+1}(t))} v_j^{(k)}(t), \quad (6.4)$$

for all $k = 1, 2, \ldots, m$.

Concerning equality constraints, we use the same idea as we did with mathematical programs. If the conditions

$$\mathbf{H}(\mathbf{u}(t), \mathbf{u}'(t)) = \mathbf{0}$$

are to be enforced, we can always do so by putting

$$g_0(\mathbf{u}(t), \mathbf{u}'(t)) \leq 0, \quad g_0(\mathbf{u}, \mathbf{p}) = \frac{1}{2} |\mathbf{H}(\mathbf{u}, \mathbf{p})|^2.$$

Even better is to set

$$\mathbf{H}(\mathbf{u}(t), \mathbf{u}'(t)) \leq \mathbf{0}, \quad -\mathbf{H}(\mathbf{u}(t), \mathbf{u}'(t)) \leq \mathbf{0}.$$

Finally, quite often, constraints come in the form of integral conditions, which are global constraints. We did see one such important example for the hanging cable. In this case, we have

$$\text{Minimize in } \mathbf{u} : \quad \int_0^T F(t, \mathbf{u}(t), \mathbf{u}'(t)) \, dt$$

under

$$\mathbf{u}(0) = \mathbf{u}_0, \mathbf{u}(T) = \mathbf{u}_T, \quad \int_0^T \mathbf{G}(t, \mathbf{u}(t), \mathbf{u}'(t)) \, dt \leq \mathbf{0}.$$

Again, we can deal with integral constraints in the form of equalities by introducing new components to $\mathbf{G}$ as we have just described.

To find good approximations for these other problems, one can apply the same technique and the same algorithm to the master functional

$$J(\mathbf{u}, \mathbf{y}) = \int_0^T F(t, \mathbf{u}(t), \mathbf{u}'(t))\, dt + \sum_{k=1}^m \exp\left(y^{(k)} \int_0^T G_k(t, \mathbf{u}(t), \mathbf{u}'(t))\, dt\right).$$

This time $\mathbf{y}$ is not a path, does not depend on the variable t, but is just a vector in $\mathbb{R}^m$. However, there is also the possibility of introducing auxiliary variables in such a way that the global integral constraints become pointwise ones. In this way, there is no need to make the distinction between both kinds of constraints, and they both can be treated within the same framework. Though specific problems may suggest better choices, in general one could introduce new components $\tilde{\mathbf{u}}$ for $\mathbf{u}$ asking for the pointwise constraints

$$\mathbf{G}(t, \mathbf{u}(t), \mathbf{u}'(t)) \leq \tilde{\mathbf{u}}'(t) \text{ for every } t \in [0, T],$$

in addition to $\tilde{\mathbf{u}}(0) = \tilde{\mathbf{u}}(T)(= \mathbf{0})$. If these conditions are met, then we can be sure that

$$\int_0^T \mathbf{G}(t, \mathbf{u}(t), \mathbf{u}'(t))\, dt \leq \mathbf{0}.$$

We will approximate in this way the standard hanging cable problem.

A convergence result similar to Theorem 4.2 can be proved in this context, though we will skip the proof as it goes beyond the scope of this elementary text. See [23] for a full treatment.

6.4   Some Examples

In this section, we plan to describe a few interesting simulations that we have conducted for a selected number of variational problems.

1. Our first example is the brachistochrone. Recall that this corresponds to a smooth, variational problem under (both) end-point restrictions, namely, the integrand is

$$F(t, u, p) = \sqrt{\frac{1 + p^2}{t}}.$$

This time both u, and p are scalar. The variational problem whose solution we would like to approximate is

$$\text{Minimize in } u: \quad I(u) = \int_0^T \sqrt{\frac{1 + u'(t)^2}{t}}\, dt$$

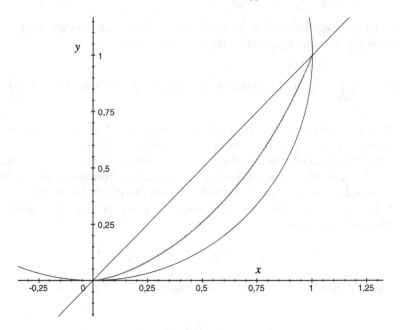

Fig. 6.1 The brachistochrone for $T = L = 1$

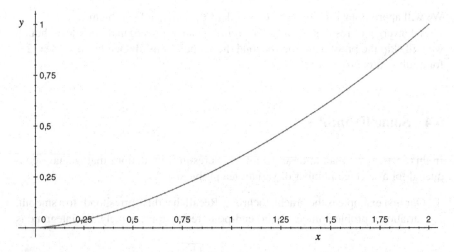

Fig. 6.2 The brachistochrone for $T = 2, L = 1$

under $u(0) = 0$, $u(T) = L$. We already pointed out that the optimal path is an arc of a cycloid. See Figs. 6.1, 6.2, and 6.3 for a picture of such an arc of a cycloid for the particular choices $T = L = 1$, $T = 2$, $L = 1$, and $T = 1$, $L = 2$, respectively. In Fig. 6.1, we have also drawn both the corresponding straight line and the circle for comparison purposes.

Fig. 6.3 The
brachistochrone for $T = 1$,
$L = 2$

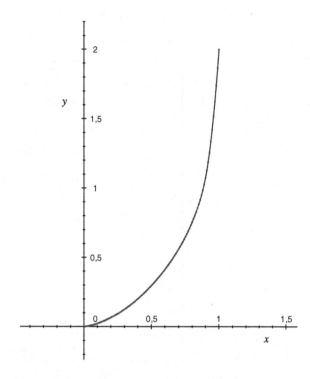

2. We now turn to another classical example where we try to find an approximation
   to the optimal profile for the variational problem

$$\text{Minimize in } u(t) : \quad \int_0^T u(t)\sqrt{1 + u'(t)^2}\, dt$$

under end-point conditions $u(0) = u_0, u(T) = u_T$. The integrand for this problem
is

$$F(t, u, p) = u\sqrt{1 + p^2}.$$

If we recall from Calculus that this is the integrand to find the surface of revolution
obtained by revolving the graph of $u(t)$ for $t \in [0, T]$ around the X-axis, then we
realize that we are seeking that particular profile furnishing the least surface area
of revolution. This is a very particular example of a much more fascinating and
difficult problem: that of the minimal surfaces. We can use our basic algorithm
to find approximations to these optimal profiles depending on the values of T,
u_0, and u_T. Figure 6.4 shows a very good approximation to the optimal profile
for the choice $T = 0.5, u_0 = 0.5, u_T = 1$.

There is a funny phenomenon with this example. In Fig. 6.5, we have drawn
the optimal profiles for the values $u_0 = u_T = 1$, but for increasing values
of $T = 1, 1.1, 1.2, 1.3, 1.33$. But if you try with $T = 1.34$ then you notice

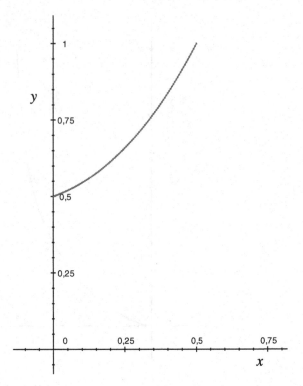

Fig. 6.4 The catenary for $T = 0.5, L = 1$

that the algorithm apparently does not converge. These difficulties were already hightlighted in the previous chapter ([20]).

3. We next focus on a typical variational problem under a global, integral constraint: the hanging cable. This is one of the main examples of a variational problem under a global, integral constraint. We seek to

$$\text{Minimize in } u(t) : \quad \int_0^1 u(t)\sqrt{1 + u'(t)^2}\, dt$$

subject to

$$u(0) = 0, u(1) = 0, \quad \int_0^1 \sqrt{1 + u'(t)^2}\, dt = L > 1.$$

As indicated earlier, we will deal with the integral constraint in a pointwise fashion by introducing a new auxiliary function $U(t)$ into the problem and demanding that

$$U'(t) = \sqrt{1 + u'(t)^2}, \quad U(0) = 0, U(1) = L.$$

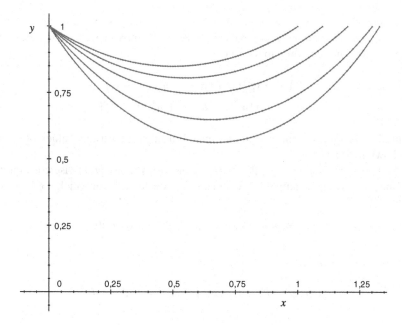

Fig. 6.5 The catenaries for $T = 1, 1.1, 1.2, 1.3, 1.33, L = 1$

Note that these conditions imply automatically that

$$\int_0^1 \sqrt{1 + u'(t)^2}\, dt = \int_0^1 U'(t)\, dt = U(1) - U(0) = L.$$

We therefore have to deal with an equivalent variational principle in two components $(u_1(t), u_2(t))$, $u_1(t) = u(t)$, $u_2(t) = U(t)$. We pretend to

$$\text{Minimize in } (u_1(t), u_2(t)): \quad \int_0^1 u_1(t)\sqrt{1 + u_1'(t)^2}\, dt$$

subject to

$$u_2'(t)^2 - u_1'(t)^2 = 1, \quad u_1(0) = u_1(1) = u_2(0) = 0, u_2(1) = L > 1. \quad (6.5)$$

Our proposal to find reasonable approximations for the optimal profiles $u = u_1$, consists in approximating the optimal solution of the free variational problem that consists in minimizing the functional

$$\int_0^1 \Big[u_1(t)\sqrt{1 + u_1'(t)^2} + \exp\big(v(t)(u_2'(t)^2 - u_1'(t)^2 - 1)\big)$$

$$+ \exp\big(w(t)(-u_2'(t)^2 + u_1'(t)^2 + 1)\big)\, \pi \Big]\, dt$$

under the given end-point conditions (6.5), for strictly positive functions $v(t)$ and $w(t)$. Once the optimal solution has been found (approximated), we update these two functions according to the rule

$$v(t) \mapsto \exp\left(v(t)(u_2'(t)^2 - u_1'(t)^2 - 1)\right)v(t),$$
$$w(t) \mapsto \exp\left(w(t)(-u_2'(t)^2 + u_1'(t)^2 + 1)\right)w(t),$$

until convergence. Optimal approximated profiles for various values of L are shown in Fig. 6.6.

4. We now turn to some typical variational problems under pointwise constraints. One such typical example is the so-called obstacle problem which corresponds to

$$\text{Minimizing in } u(t): \quad \int_0^T \frac{1}{2}|u(t)'|^2\,dt$$

under the constraints

$$u(0) = u(T) = 0, \quad u(t) \geq \phi(t) \text{ for all } t \in [0, T].$$

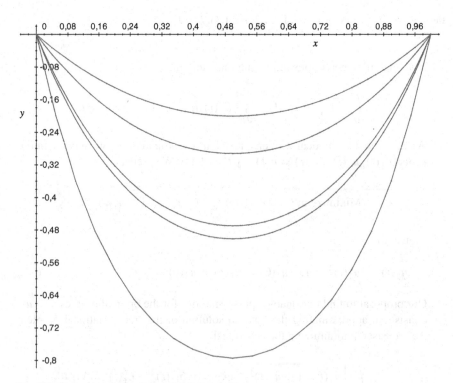

Fig. 6.6 The catenaries for $L = 1.1, 1.2, 1.4, 1.45, 2$

The given function $\phi(t)$ provides the profile for the obstacle, while feasible $u(t)$ tries to minimize the elastic energy of a certain string clamped in its two endpoints, that cannot pass through the obstacle. According to our discussion in Sect. 6.3, the numerical procedure is now an iterative process where each iteration includes two steps:

(a) find the optimal solution of the regular variational problem

$$\text{Minimize in } u(t): \quad \int_0^T \left[\frac{1}{2}|u'(t)|^2 + e^{v(t)(\phi(t)-u(t))} \right] dt$$

under $u(0) = u(T) = 0$, for an auxiliary given function $v(t)$;

(b) update rule for this auxiliary function

$$v(t) \mapsto v(t)e^{v(t)(\phi(t)-u(t))}.$$

One should go over this two-step process until convergence.

Figures 6.7, 6.8, and 6.9 show three such approximations for a symmetric obstacle, a non-symmetric one, and a non-convex one, respectively.

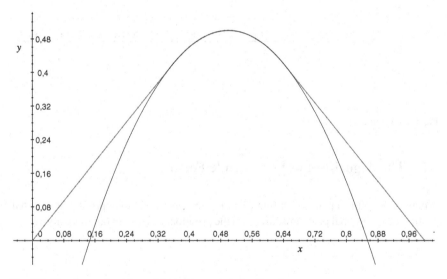

Fig. 6.7 The shape of a string subjected to an obstacle, $T = 1$

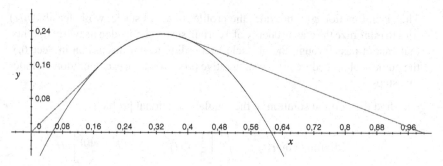

Fig. 6.8 A non-symetric obstacle

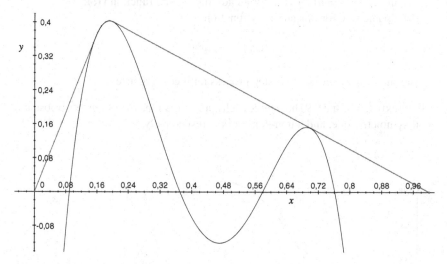

Fig. 6.9 A non-convex obstacle

6.5   The Algorithm in Pseudocode Form

It may be helpful to readers willing to build their own code to approximate optimal profiles of variational problems, to rely in the pseudocode form of our main numerical procedure.

We focus on a generic problem

$$\text{Minimize in } \mathbf{u}: \quad \int_0^T F(t, \mathbf{u}(t), \mathbf{u}'(t))\, dt$$

under

$$\mathbf{u}(0) = \mathbf{u}_0, \mathbf{u}(T) = \mathbf{u}_T, \quad \mathbf{G}(\mathbf{u}(t), \mathbf{u}'(t)) \le 0 \text{ for all } t \in [0, T].$$

As indicated earlier $\mathbf{u}_0, \mathbf{u}_T$ are given vectors in $\mathbb{R}^N$, while

$$F(t, \mathbf{u}, \mathbf{p}) : [0, T] \times \mathbb{R}^N \times \mathbb{R}^N \to \mathbb{R}, \quad \mathbf{G}(\mathbf{u}, \mathbf{p}) : \mathbb{R}^N \times \mathbb{R}^N \to \mathbb{R}^m$$

are the cost integrand, and the mapping to set up constraints, respectively.

The various steps to cover are:

- Input dimensions N, m; end-points $T_0 = 0$, $T_1 = T$; discretization parameter n so that $h = (T_1 - T_0)/n$ is the distance between adjacent nodes; and end-point vectors $\mathbf{u}_0, \mathbf{u}_1$.
- Input initial iterate $\mathbf{u}$ as the affine map taking on the values $\mathbf{u}_0, \mathbf{u}_1$ at $t = T_0, t = T_1$, respectively. It is a $(n + 1) \times N$ matrix.
- Input multiplier v as a non-negative, $m \times n$-component vector: one m-vector for each of the n subintervals of the discretization of the interval $[T_0, T_1]$.
- Input threshold value for stopping criterium ε.
- Main iterative step.

  - Stopping criterium. Compute the discrete derivative of $\mathbf{u}$, and its average value over each of the subintervals of the discretization, and calculate the value of $\mathbf{G}$ over each such subinterval. If the individual, componentwise products $\mathbf{G} \times \mathbf{v}$ in all of the subintervals of the discretization is smaller than ε, stop and take the current values of $\mathbf{u}$ as a good approximation of the optimal profile of the problem. If some of those values are greater than ε, proceed.
  - Main iterative loop to update $\mathbf{u}$ and $\mathbf{v}$.
    Starting from the current approximation $\mathbf{u}$, use formulas for the steepest descent and step size as explained earlier in the chapter for the augmented functional as in (6.3)

$$J(\mathbf{u}, \mathbf{v}_j) = \int_0^T \left[F(t, \mathbf{u}(t), \mathbf{u}'(t)) + \sum_{k=1}^m e^{v_j^{(k)}(t) G_k(\mathbf{u}(t), \mathbf{u}'(t))} \right] dt,$$

until no further improvement measured by ε is possible.
Update $\mathbf{v}$ according to (6.4)

$$v_{j+1}^{(k)}(t) = e^{v_j^{(k)}(t) G_k(\mathbf{u}_{j+1}(t), \mathbf{u}'_{j+1}(t))} v_j^{(k)}(t).$$

The main part of the code consisting in approximating the optimal profile of the augmented functional under no constraint (except end-point conditions) may be borrowed from other sources (like MatLab).

6.6 Exercises

6.6.1 Exercises to Support the Main Concepts

1. Look at the variational problem

$$\text{Minimize in } u(x) : \quad \int_0^1 \left[\frac{1}{2} u'(x)^2 + \frac{1}{2} (u(x) - 10)^2 \right] dx$$

   subject to

$$u(0) = u(1) = 0, \quad u(x) \leq 1.$$

   (a) By neglecting the pointwise constraint $u(x) \leq 1$, show that it has to be active in a certain subinterval of $[0, 1]$.
   (b) To approximate numerically the optimal solution, we need to examine the modified variational principle

$$\text{Minimize in } u(x) : \quad \int_0^1 \left[\frac{1}{2} u'(x)^2 + \frac{1}{2} (u(x) - 10)^2 + \exp(v(x)(u(x) - 1)) \right] dx$$

   under the same end-point conditions for each given $v(x) > 0$. Check if this new problem admits optimal solutions for any such $v(x)$. Is it unique?
   (c) Find an approximation of the optimal solution.

2. For the variational problem

$$\text{Minimize in } u(x) : \quad \int_{-1}^1 \frac{1}{2} u'(x)^2 \, dx$$

   under

$$u(-1) = u(1) = 0, \quad u(x) \geq x^3 - 2x - 2x^2,$$

   argue that the inequality has to be active somewhere in $[-1, 1]$, that the modified variational principle for the main step of the iterative process is well-posed (it has a unique optimal solution), and calculate an approximation of the optimal solution.

3. For the integral cost

$$\int_0^1 (x'(t)^2 + x(t)^2) \, dt$$

   the best piecewise-linear, continuous function with $x(0) = 0$, $x(1) = 1$, and linear in $[0, 1/2]$, and in $[1/2, 1]$ (with different slopes) is to be sought.

(a) Write in an explicit way those functions in the form

$$x(t) = \begin{cases} \text{linear expression, } 0 < t < 1/2, \\ \text{linear expression, } 1/2 < t < 1, \end{cases}$$

in terms of the value u at the node $1/2$. Recall that $x(0) = 0$, $x(1) = 1$.

(b) By plugging this function into the cost, write a new objective function in terms of u.

(c) Find the optimal value u (and so, the best such piecewise-linear, continuous function), by minimizing the cost of the previous item. Compare such optimal value u with the value at $t = 1/2$ of the true minimizer.

6.6.2 Computer Exercises

1. The variational problem

$$\text{Minimize in } u(x): \quad \int_0^1 (u(x)^2 - xu(x) + 3)\,dx$$

under no constraints was found to be $u(x) = x/2$ in the last chapter. Approximate the optimal solution of the family of variational problems

$$\text{Minimize in } u(x): \quad \int_0^1 \left[\frac{\varepsilon}{2}u'(x)^2 + (u(x)^2 - xu(x) + 3)\right] dx$$

under $u(0) = 1/2$, $u(1) = 0$, for decreasing values of ε, always kept positive. What is the behavior you observe at the end-points as ε becomes smaller?

2. Explore numerically the variational problem

$$\text{Minimize in } u(t): \quad \int_0^1 [tu'(t) + u(t)^2]\,dt$$

under $u(0) = 0$, $u(1) = 1$. Can you conclude anything?

3. For increasing values of $n = 1, 2, 3, \ldots$, find approximations of the optimal solutions of the variational problems

$$\text{Minimize in } u(x): \quad \int_0^1 \left[u'(x)^2 + e^{nu(x)}\right] dx$$

under $u(0) = 0$, $u(1) = 1$.

4. Find an approximation of the optimal solution of the constrained variational problem

$$\text{Minimize in } y(x): \quad \int_0^1 y'(x)^2 \, dx$$

subject to

$$y(0) = y(1) = 0, \quad \int_0^1 y(x) \, dx = 1.$$

(a) By introducing a new variable $z'(x) = y(x)$, reduce the problem to a second-order variational problem, whose solution can be easily found.
(b) Setup the approximation of the problem in the two unknowns (z, y). Note that the approximation does not seem to work smoothly, and is quite sensitive to the value of the multipliers used as initialization. Do you find some explanation for this behavior?
(c) Is there any significant change if we replace the cost functional by

$$\int_0^1 y'(x)^4 \, dx$$

instead? Find the optimal profile.

5. For the family of functionals

$$\int_0^1 \left[\frac{1}{p} u'(x)^p + u(x) \right] dx,$$

find the minimizer under $u(0) = 0$, $u(1) = 1$ depending on p. Perform an approximation of such optimal solution for $p = 2$, $p = 4$, and $p = 8$, and compare your results.

6. An optimal steering problem. Set $\mathbf{x} = (x_1, x_2, x_3)$ and $\mathbf{x}_1 = (0, 0, a)$, $a > 0$. Consider the problem

$$\text{Minimize in } \mathbf{x}: \quad \int_0^1 \frac{1}{2} (|x_1'(t)|^2 + |x_2'(t)|^2) \, dt$$

subject to

$$\mathbf{x}(0) = \mathbf{0}, \mathbf{x}(1) = \mathbf{x}_1, \quad x_3'(t) = x_2(t)x_1'(t) - x_1(t)x_2'(t) \text{ for all } t.$$

(a) Try to find an approximation of the optimal curve by using as a starting path the line segment joining the initial and final points. Take $a = 1$. Can you guess a reason for this behavior?
(b) Use a different path as a starting path for the iterations, and find an approximation of the (presumed) optimal path.

6.6.3 Case Studies

1. Constrained mechanical systems. If $L > 0$ is given, the variational problem

$$\text{Minimize in } \mathbf{x} : \quad \int_0^1 \left[\frac{1}{2}|\mathbf{x}'(t)|^2 - g x_2(t) \right] dt$$

for $\mathbf{x} = (x_1, x_2)$, under

$$\mathbf{x}(0) = \mathbf{x}_0, \mathbf{x}(1) = \mathbf{x}_1, \quad |\mathbf{x}(t)|^2 = L^2 \text{ for every } t,$$

accounts for the movement of a particle under the action of gravity g but restricted to moving in a circle. Find an approximation of several such situations for different initial and/or final conditions.

2. Any dynamical system

$$\mathbf{x}'(t) = \mathbf{F}(\mathbf{x}(t)) \text{ in } (0, T), \quad \mathbf{x}(0) = \mathbf{x}_0$$

can be looked at by realizing that integral curves are absolute minimizers of the error functional

$$E(\mathbf{x}) = \int_0^T \frac{1}{2}|\mathbf{x}'(t) - \mathbf{F}(\mathbf{x}(t))|^2 dt$$

under an initial condition $\mathbf{x}(0) = \mathbf{x}_0$. In this way, approximation of integral curves of the system can be found with the ideas in this chapter. For the standard harmonic oscillator, take

$$\mathbf{F}(\mathbf{x}) = \begin{pmatrix} 0 & 1 \\ -1 & 0 \end{pmatrix} \mathbf{x}, \quad \mathbf{x} = (x_1, x_2)$$

and $\mathbf{x}_0 = (1, 0)$, and compute the corresponding integral curves for increasing values of T. Select another field $\mathbf{F}$, and calculate some approximations to several integral curves.

3. An obstacle problem. Suppose a elastic string, fixed at its two end-points, holds a circular solid in equilibrium. If $u(x)$ is the profile of the string with $u(-1) = u(1) = 0$, and the circular solid is represented by the equation

$$v(x)^2 + x^2 = 1/4,$$

then $u(x)$ is the result of the minimization process

$$\text{Minimize in } u(x) : \quad \int_{-1}^1 \frac{1}{2} u'(x)^2 dx$$

under

$$u(-1) = u(1) = 0, \quad u(x) \le -\sqrt{0.25 - x^2} \text{ for } x \in [-0.5, 0.5].$$

Approximate such optimal profile $u(x)$. What if the obstacle is not located symmetrically with respect to the string

$$u(x) \le -\sqrt{0.25 - (x - 0.2)^2} \text{ for } x \in [-0.3, 0.8]?$$

4. Newton's solid of revolution.[1] If a radially-shaped solid is moving, with constant velocity in a straight line, through a fluid under some specific conditions, we wonder about the optimal shape of such a solid which offers minimum resistance to movement. It turns out that the optimal profile must minimize the functional

$$\int_0^1 \frac{x}{1 + y'(x)^2} \, dx$$

under the conditions

$$y(1) = 0, \quad y'(x) \le 0, \quad y(x) \le M,$$

for some constant M.

(a) Check that the integrand is not convex on the derivative variable.
(b) Go about the numerical approximation of the optimal profile in the following way. Given the radial symmetry, set up the problem through minimization of the functional

$$\int_{-1}^1 \frac{|x|}{1 + y'(x)^2} \, dx$$

under $y(-1) = y(1) = 0, 0 \le y(x) \le M$. Take $M = 1, 2, 3, 4$. The optimal curve will show a "hole" around the origen. If you flatten this hole you will get a good approximation of the optimal profiles for those values of M.

References

1. H. Attouch, G. Buttazzo, G. Michaille, *Variational Analysis in Sobolev and BV Spaces. Applications to PDEs and Optimization*, MPS/SIAM Series on Optimization. Society for Industrial and Applied Mathematics (SIAM), Mathematical Programming Society (MPS) vol. 6 (Philadelphia, PA, 2006)
2. B. van Brunt, *The Calculus of Variations*, Universitext (Springer-Verlag, New York, 2004)

---

[1]See, for instance among many other references, [25].

3. J.A. Burns, *Introduction to the Calculus of Variations and Control with Modern Applications*, Chapman & Hall/CRC Applied Mathematics and Nonlinear Science Series (CRC Press, Boca Raton, FL, 2014)

4. G. Buttazzo, M. Giaquinta, S. Hildebrandt, *One-dimensional Variational Problems. An introduction*, Oxford Lecture Series in Mathematics and its Applications, vol. 15 (The Clarendon Press, Oxford University Press, New York, 1998)

5. K. Ch. Chang, *Lecture Notes on Calculus of Variations*, translated by Tan Zhang. Peking University Series in Mathematics, vol. 6 (World Scientific Publishing Co. Pte. Ltd., Hackensack, NJ, 2017)

6. F. Clarke, *Functional Analysis, Calculus of Variations and Optimal Control*, Graduate Texts in Mathematics, vol. 264 (Springer, London, 2013)

7. B. Dacorogna, *Direct Methods in the Calculus of Variations*, 2nd edn. Applied Mathematical Sciences, vol. 78 (Springer, New York, 2008)

8. B. Dacorogna, *Introduction to the Calculus of Variations*, 2nd edn. (World Scientific, 2008)

9. I. Ekeland, R. Tmam, *Convex Analysis and Variational Problems*, translated from the French. Corrected reprint of the 1976 English edition. Classics in Applied Mathematics. Society for Industrial and Applied Mathematics (SIAM), vol. 28 (Philadelphia, PA, 1999)

10. P. Freguglia, M. Giaquinta, *The Early Period of the Calculus of Variations*, (Birkhuser/Springer, [Cham], 2016)

11. J.B. Hiriart-Urruty, C. Lemarchal, *Convex Analysis and Minimization Algorithms. I. Fundamentals*, Grundlehren der Mathematischen Wissenschaften [Fundamental Principles of Mathematical Sciences], vol. 305 (Springer-Verlag, Berlin, 1993)

12. A.O. Ivanov, A.A. Tuzhilin, *Branching Solutions to One-dimensional Variational Problems* (World Scientific Publishing Co., Inc, River Edge, NJ, 2001)

13. L. Komzsik, *Applied Calculus of Variations for Engineers*, 2nd edn. (CRC Press, Boca Raton, FL, 2014)

14. M. Kot, *A First Course in the Calculus of Variations, Student Mathematical Library*, vol. 72 (American Mathematical Society, Providence, RI, 2014)

15. C. Lanczos, *The Variational Principles of Mechanics* (Dover Publications, INC, New York, 1970)

16. M. Levi, *Classical Mechanics with Calculus of Variations and Optimal Control. An Intuitive Introduction*, Student Mathematical Library. American Mathematical Society, Providence, RI; Mathematics Advanced Study Semesters, vol. 69 (University Park, PA, 2014)

17. D. Liberzon, *Calculus of Variations and Optimal Control Theory*, A Concise Introduction (Princeton University Press, Princeton, NJ, 2012)

18. Ch. R. MacCluer, *Calculus of Variations. Mechanics, Control, and other Applications* (Pearson Prentice Hall, Upper Saddle River, NJ, 2005)

19. M. Mesterton-Gibbons, *A Primer on the Calculus of Variations and Optimal Control Theory*, Student Mathematical Library, vol. 50 (American Mathematical Society, Providence, RI, 2009)

20. J. Oprea, *The Mathematics of Soap Films. Explorations with Maple*, vol. 10, (Student Math. Library, AMS, 2000)

21. P. Pedregal, *Introduction to Optimization*, Texts Appl. Math. vol. 46 (Springer, 2003)

22. P. Pedregal, On the Generality of Variational Principles. Milan J. Math. **71**, 319–356 (2003)

23. P. Pedregal, A Direct Numerical Algorithm for Constrained Variational Problems. Numer. Func. Anal. Opt. **38**(4), 486–506 (2017)

24. E.R. Pinch, *Optimal Control and the Calculus of Variations*, Oxford Science Publications (The Clarendon Press, Oxford University Press, New York, 1993)

25. Q. Hui, Newton's Minimal Resistance Problem. Proj. Math. **7581**, April 13, (2006)

26. A. Takayama, *Mathematical Economics*, 2nd edn. (Cambridge University Press, Cambridge, 1985)

27. J.L. Troutman, *Variational Calculus and Optimal Control. With the Assistance of William Hrusa, Optimization with elementary convexity*, 2nd edn. Undergraduate Texts in Mathematics. (Springer-Verlag, New York, 1996)

28. F.Y.M. Wan, *Introduction to the Calculus of Variations and its Applications*, Chapman and Hall Mathematics Series (Chapman & Hall, New York, 1995)

Part III
Optimal Control

Chapter 7
Basic Facts About Optimal Control

Just as variational problems, optimal control problems are optimization problems where possibilities and strategies are represented by functions, and costs are expressed, most of the time, by integrals. But there are some distinctive features of optimal control problems that make them particularly interesting for many situations in science and engineering. Probably the most important ingredient in optimal control is the presence of the state law through which we are entitled to interact with the system to our advantage. As usual, we will first present some selected examples to understand the scope and the nature of these optimization problems; we will then move on to describe a typical model problem and its main variants, to finally focus on the use and understanding of optimality conditions through Pontryaguin's principle.

7.1 Some Motivating Examples

Example 7.1 A farmer earns a living by selling the honey produced by bees. If at a given time t, the density of the population of bees in the hives is given by $x(t)$, then it is known that the law governing the growth of it is $x'(t) = x(t)$. Suppose that at an initial time $t = 0$, the farmer measures $x(0) = 1/4$, and he sets to himself an horizon $T = 1$, when he will sell the available honey at a unit price $2/e$. He has the possibility of introducing new individuals in the hives so as to increase the growth ratio. This possibility is modeled by $u(t)$ with $0 \le u(t) \le 1$, in such a way that the growth law becomes $x'(t) = x(t) + u(t)$, but the farmer incurs in a cost given by the integral

$$\int_0^1 \frac{1}{2} u(t)^2 \, dt.$$

The issue is to recommend the optimal strategy $u(t)$ so as to maximize net income at the end of the period.

© Springer International Publishing AG 2017
P. Pedregal, *Optimization and Approximation*, UNITEXT 108,
DOI 10.1007/978-3-319-64843-9_7

Example 7.2 A certain spacecraft, departing from rest, is to be positioned at a distance $3 + 5/6$ units from the starting point (in a straight line) in $T = 3$ time units. We would like to do so at a minimal cost associated with the use of the accelerator $u(t)$, under the restriction $0 \leq u(t) \leq 1$, where $x''(t) = u(t)$, so that $x(t)$ indicates the location, with respect to the rest position, of the spacecraft at time t, and the cost is measured by the integral

$$\int_0^3 ku(t)^2 \, dt.$$

k is a given, positive parameter.

These two examples are quite good representatives of optimal control problems. Though they are pretty elementary, they make us comprehend that we are before an important area of the application of mathematics. Let us dwell a bit more on each example as to better understand and appreciate the amazing conclusion that is meant by being able to find optimal strategies.

1. In the farmer example, we have at our disposal all strategies $u(t)$ provided $0 \leq u(t) \leq 1$ for $t \in [0, 1]$. We decide what to do at each instant of the full horizon interval. Each such decision will have a corresponding net income at the end. For example, we may decide not to do anything so that $u(t) \equiv 0$. In that case, the growth of the bee population will evolve according to the law $x'(t) = x(t)$ with $x(0) = 1/4$. We know from the basic theory of ODE that the solution of this problem is $x(t) = (1/4)e^t$, and so the amount of honey at the end of the period would be $x(1) = e/4$. Since we have not spent any additional money, the net income would be $1/2$.

   On the other side of the spectrum, we find the possibility of introducing new individuals at maximum rate $u(t) \equiv 1$. We will definitely have much more honey at the end, but we have to pay a price for doing that which is given by the above integral. In this case, this additional expense is $1/2$. In order to find the amount of honey for $t = 1$, we need to solve the problem

$$x'(t) = x(t) + 1(u \equiv 1) \text{ in } (0, 1), \quad x(0) = 1/4.$$

   The solution is $x(t) = -1 + (5/4)e^t$, and $x(1) = (5/4)e - 1$. Net income is $2(1 - 1/e)$, which is greater than the previous one. From these two easy calculations, we find that the strategy $u \equiv 1$ is better (net income is bigger) than $u \equiv 0$. But there are infinitely many intermediate strategies because we are entitled to change our strategy at every time as long as $0 \leq u(t) \leq 1$. Feasible strategies would be $u(t) = \cos^2 t$, $u(t) = t^2$ $(0 \leq t \leq 1)$, $u(t) = 1/2$, $u(t) = 1/(1 + t^2)$, etc. The issue is whether there is any possible way to find the best one, obviously without an exhaustive analysis of all possibilities, which would be, of course, impossible.

2. There are some interesting points to be made about the second example. We have that $x(0) = x'(0) = 0$ because we are informed that the spacecraft departs from rest. If we choose not to use the accelerator $u \equiv 0$, then nothing happens, and the spacecraft does not move. We would not have to pay anything, but we can hardly be at a different place after $T = 3$ units of time. Suppose we accelerate at maximum rate at all times $u \equiv 1$. We will have to pay a price $3k$, but this time the spacecraft has definitely moved. Where will we be after $T = 3$ time units? We need to solve the problem

$$x''(t) = 1 \text{ in } (0, 3), \quad x(0) = x'(0) = 0.$$

The solution is very easily found to be $x(t) = (1/2)t^2$, and so $x(3) = 9/2$ which is clearly greater than $3 + 5/6$. We have over-accelerated, and passed through the point $3 + 5/6$ before $T = 3$. This means that we could have saved some fuel, and yet be at the indicated point, at the required time. Again, is there a way to find the best (cheapest) use one can make of the accelerator to meet the requirements imposed?

The whole point of optimal control is to learn how to find and/or approximate those optimal strategies.

The next example is important for its relevance in Engineering, though we do not specify concrete examples.

Example 7.3 A general mechanical system is susceptible of being controlled. As a matter of fact, this is one of the most appealing applications of control theory. If vector $\mathbf{q}$ incorporates all of the angles at the various joints of the system (a robot, let us say), then the controlled mechanics can be written in the general form

$$H(\mathbf{q})\mathbf{q}'' + \left[\frac{1}{2}H(\mathbf{q})' + S(\mathbf{q}, \mathbf{q}')\right]\mathbf{q}' + \mathbf{g}(\mathbf{q}) = \mathbf{u},$$

where

- $\mathbf{q}'$ is the vector of angular velocities, and $\mathbf{q}''$ the vector of angular accelerations;
- $H(\mathbf{q})$ is the matrix of inertia, and $g(\mathbf{q})$ is the gravity torque vector (gradient of the gravity potential);
- $S(\mathbf{q}, \mathbf{q}')$ is a certain skew symmetric tensor related to the inertia matrix;
- $\mathbf{u}$, the control, is the applied (externally) torque at each joint.

The optimal control problem might consist in finding the optimal use of the control variable to adopt a certain desirable configuration $\mathbf{q}_T$ in minimal time, starting from another given one $\mathbf{q}_0$. Many other possibilities for the cost functional could be explored.

7.2 An Easy Existence Result

After the analysis we have performed with those two previous examples, it is not difficult to grasp the structure of a typical model problem of optimal control. We seek to

Minimize in

$$\mathbf{u}(t): \quad I(\mathbf{u}) = \int_0^T F(t, \mathbf{x}(t), \mathbf{u}(t)) \, dt \, (+\phi(\mathbf{x}(T)))$$

subject to

$$\mathbf{x}'(t) = \mathbf{f}(t, \mathbf{x}(t), \mathbf{u}(t)) \text{ in } (0, T), \quad \mathbf{x}(0) = \mathbf{x}_0, (\mathbf{x}(T) = \mathbf{x}_T),$$

$$\mathbf{x}(t) \in \Omega \subset \mathbb{R}^N, \quad \mathbf{u}(t) \in \mathbf{K} \subset \mathbb{R}^m.$$

Let us explain with some care the various ingredients, and clarify notation.

- $T > 0$ is the time horizon considered. It can take any positive value, and even $+\infty$. Sometimes, it may not be specified, and finding it, in an optimal way, is part of the problem.
- The set of variables $\mathbf{x}(t) : [0, T] \to \mathbb{R}^N$ are the state variables. It is a time-dependent vector valued function. It indicates where the system we are concerned about stands for each time. Specific values of those N parameters determine completely the state of the system.
- The set of variables $\mathbf{u}(t) : [0, T] \to \mathbb{R}^m$ are the control variables. Through them, we interact, manipulate, and control the system to our advantage. They express all of the various strategies at our disposal.
- Vectors $\mathbf{x}_0 \in \mathbb{R}^N$ and $\mathbf{x}_T \in \mathbb{R}^N$ specify where the system stands when we start to care about it, and, if applicable, where we would like the system to end. We have written the final condition $\mathbf{x}(T) = \mathbf{x}_T$ within parentheses to indicate that, depending on the particular situation, we will or will not have to enforce this final condition.
- The state law $\mathbf{f}(t, \mathbf{x}, \mathbf{u}) : [0, T] \times \mathbb{R}^N \times \mathbb{R}^m \to \mathbb{R}^N$ governs the evolution of the system with time once we have decided a specific strategy $\mathbf{u}(t)$. Indeed, the differential system of equations

$$\mathbf{x}'(t) = \mathbf{f}(t, \mathbf{x}(t), \mathbf{u}(t)) \text{ in } (0, T), \quad \mathbf{x}(0) = \mathbf{x}_0,$$

determines in a unique way how the system will evolve starting out from $\mathbf{x}_0 \in \mathbb{R}^N$.
- Constraints for both sets of variables. $\Omega \subset \mathbb{R}^N$ is the set where state variables are allowed, to avoid the system to run into undesirable regimes. Likewise $\mathbf{K} \subset \mathbb{R}^m$

is the feasible set for the control variable. This constraint takes into account, in practice, that we cannot force the system beyond reasonable limits without risking collapse.

- The cost functional is given through a time integral over the full time interval $[0, T]$ with an integrand $F(t, \mathbf{x}, \mathbf{u}) : [0, T] \times \mathbb{R}^N \times \mathbb{R}^m \to \mathbb{R}$. It can incorporate an additional term with a function $\phi : \mathbb{R}^N \to \mathbb{R}$ depending on where the system has finished. This second contribution does not occur when a final condition $\mathbf{x}(T) = \mathbf{x}_T$ is to be respected. The role of the cost functional is crucial. It serves to compare various strategies, and decide which one is better because the price of one is less than the price of the others.

The way in which the control problem works is the following. The goal is to find the optimal strategy $\overline{\mathbf{u}}(t)$ among those $\mathbf{u}(t)$ that have been declared feasible. To begin with, we require that $\mathbf{u}(t) \in \mathbf{K}$ for all times. Every such strategy has an associated cost that is calculated as follows. Solve first the problem

$$\mathbf{x}'(t) = \mathbf{f}(t, \mathbf{x}(t), \mathbf{u}(t)) \text{ in } (0, T), \quad \mathbf{x}(0) = \mathbf{x}_0.$$

This is an initial-value problem for $\mathbf{x}$, once we have chosen $\mathbf{u}$, which, typically, admits a unique solution. If the final condition $\mathbf{x}(T) = \mathbf{x}_T$ is to be enforced, then not all possibilities $\mathbf{u}(t) \in \mathbf{K}$ will lead to it. So this final condition, indirectly, also imposes a restriction on the feasible strategies. In addition, if the condition $\mathbf{x}(t) \in \Omega$ has to be respected, it also imposes quite tight restrictions on the control as well. These are much harder to deal with, and so we will ignore them altogether: Ω will always be taken to be all of space $\mathbb{R}^N$. Once we have found $\mathbf{x}(t)$, then, together with its mate $\mathbf{u}(t)$, we can finally compute the cost of $\mathbf{u}$ by calculating

$$I(\mathbf{u}) = \int_0^T F(t, \mathbf{x}(t), \mathbf{u}(t)) \, dt (+\phi(\mathbf{x}(t))).$$

Definition 7.1 A feasible strategy $\overline{\mathbf{u}}$ is optimal for our problem, if $I(\overline{\mathbf{u}}) \leq I(\mathbf{u})$ for every possible admissible control $\mathbf{u}$.

An important issue to be examined before one studies optimality conditions is that of the existence of optimal strategies. This is rather meaningful, because if a particular problem, for whatever reason, does not admit optimal solutions, there is no point in trying to use optimality conditions to find an object which does not exist. We state one main such existence theorem without proof, just for the sake of completeness. It can be used to ensure existence of optimal solutions in all of the examples that we will be looking at.

Theorem 7.1 *Suppose the following conditions hold:*

1. *the cost integrand $F(t, \mathbf{x}, \mathbf{u})$ is a convex function regarded as a function of pairs $(\mathbf{x}, \mathbf{u})$ for every time t, and coercive in the sense*

$$\lim_{|\mathbf{u}| \to \infty} \frac{F(t, \mathbf{x}, \mathbf{u})}{|\mathbf{u}|} = +\infty,$$

uniformly with respect to every time t, and every vector $\mathbf{x}$;
2. *the mapping $\mathbf{f}(t, \mathbf{x}, \mathbf{u})$ is a linear mapping regarded as a mapping of pairs $(\mathbf{x}, \mathbf{u})$ for every time t;*
3. *the set $\mathbf{K}$ is closed and convex.*

Then our basic optimal control problem (taking $\Omega = \mathbb{R}^N$) admits an optimal solution.

Though we have described a pretty general framework for optimal control problems governed by ODEs, our aim in this textbook is quite modest. In most of our examples dimensions N and m will be kept as low as possible: $N = m = 1$. In some examples $N = 2$, but we will always have $m = 1$. In such situations, $\mathbf{K}$ will usually be an interval, possibly infinite. Moreover, the map $\mathbf{f}$ determining the state law will always be taken to be linear in $(\mathbf{x}, \mathbf{u})$ (as Theorem 7.1 requires), and autonomous with no explicit dependence on time. Similarly for the functions involved in the cost functional: F will not depend upon time, and will depend at most quadratically on the control and/or the state. Whenever present, ϕ will at most be quadratic as well. This particular class of optimal control problems are important in Engineering, and are identified as LQR (Linear Quadratic Regulator) problems.

In understanding Pontryaguin's maximum principle, which is the universal term utilized to refer to optimality conditions for optimal control problems, we will proceed in three steps:

1. We will first treat the case where there is no ϕ, and there is no restriction set $\mathbf{K}$.
2. We will then focus on the situation where we have a finite restriction set $\mathbf{K}$, but still with no ϕ.
3. The case for a non-trivial ϕ will be explored.

7.3 Optimality Conditions

We start looking at optimality for problems of the form

$$\text{Minimize in } \mathbf{u}(t): \quad I(\mathbf{u}) = \int_0^T F(\mathbf{x}(t), \mathbf{u}(t)) \, dt$$

under
$$\mathbf{x}'(t) = \mathbf{f}(\mathbf{x}(t), \mathbf{u}(t)) \text{ in } (0, T), \quad \mathbf{x}(0) = \mathbf{x}_0.$$

In this neat, simple form, we can explain where optimality conditions come from, and why they have the form they do. Other important ingredients make the discussion more involved, and well beyond the scope of this text. In successive forms of optimal control, we will be contented with understanding and manipulating optimality conditions for simple cases.

Suppose $\mathbf{u}$ is an optimal solution of our optimal control written at the beginning of this section. Let $\mathbf{U}$ be a variation of $\mathbf{u}$. By this we simply mean that $\mathbf{U}$ is used to perturb the optimal solution $\mathbf{u}$ in the form $\mathbf{u} + \varepsilon\mathbf{U}$ for a small parameter ε. If we were to enforce further constraints on controls, then the combination $\mathbf{u} + \varepsilon\mathbf{U}$ might not be feasible without some additional requirement. But in this general situation, it is. This perturbation $\mathbf{u} + \varepsilon\mathbf{U}$ of the optimal solution will, most definitely, produce a certain perturbation on the associated state through the state law. If we let $\mathbf{X}$ be such a perturbation, then we should have

$$\mathbf{x}'(t) + \varepsilon\mathbf{X}'(t) = \mathbf{f}(\mathbf{x}(t) + \varepsilon\mathbf{X}(t), \mathbf{u}(t) + \varepsilon\mathbf{U}(t)) \text{ in } (0, T).$$

To simplify a bit notation, we will write

$$\mathbf{x}' + \varepsilon\mathbf{X}' = \mathbf{f}(\mathbf{x} + \varepsilon\mathbf{X}, \mathbf{u} + \varepsilon\mathbf{U}).$$

Because the two variables t and ε are independent, we can differentiate with respect to ε this last equation, and then set $\varepsilon = 0$, to find

$$\mathbf{X}' = \mathbf{f}_\mathbf{x}(\mathbf{x}, \mathbf{u})\mathbf{X} + \mathbf{f}_\mathbf{u}(\mathbf{x}, \mathbf{u})\mathbf{U}.$$

All products between bold-faced letter objects are always taken to be vector or matrix multiplication, as it is appropriate in every situation, without further comment. Since the optimal pair $(\mathbf{x}, \mathbf{u})$ is known, we can regard this last linear system as yielding the variation $\mathbf{X}$ of $\mathbf{x}$ produced by the perturbation $\mathbf{U}$ of $\mathbf{u}$, provided we require $\mathbf{X}(0) = \mathbf{0}$ to ensure that perturbations $\mathbf{x} + \varepsilon\mathbf{X}$ do not change the initial value $\mathbf{x}_0$. Likewise, the local change in the cost procuded by $\mathbf{U}$ will be measured by the integral

$$\int_0^T [F_\mathbf{x}(\mathbf{x}, \mathbf{u})\mathbf{X} + F_\mathbf{u}(\mathbf{x}, \mathbf{u})\mathbf{U}]\, dt. \tag{7.1}$$

We would like to transform this change in (7.1) so that only the perturbation $\mathbf{U}$, and not $\mathbf{X}$, occurs in it so as to find a neat condition for the pair $(\mathbf{x}, \mathbf{u})$ to be optimal. We need to take advantage of the differential system that $\mathbf{X}$ satisfies in order to eliminate $\mathbf{X}$ in the variation of the cost. At this stage the costate (or multiplier) $\mathbf{p}(t)$ is introduced. It is the solution of the system

$$-\mathbf{p}'(t) = \mathbf{p}(t)\mathbf{f}_\mathbf{x}(\mathbf{x}(t), \mathbf{u}(t)) + F_\mathbf{x}(\mathbf{x}(t), \mathbf{u}(t)) \text{ in } (0, T), \quad \mathbf{p}(T) = \mathbf{0}. \tag{7.2}$$

We will immediately understand the reason for this new system. Let us focus on the first term of (7.1). Bearing in mind (7.2), we can write

$$\int_0^T F_{\mathbf{x}}(\mathbf{x}, \mathbf{u})\mathbf{X}\,dt = -\int_0^T [\mathbf{p}'\mathbf{X} + \mathbf{p}\mathbf{f}_{\mathbf{x}}(\mathbf{x}, \mathbf{u})\mathbf{X}]\,dt.$$

If we integrate by parts in the first term of the right-hand side, because of the end-points conditions $\mathbf{X}(0) = \mathbf{p}(T) = \mathbf{0}$,

$$\int_0^T F_{\mathbf{x}}(\mathbf{x}, \mathbf{u})\mathbf{X}\,dt = \int_0^T [\mathbf{p}\mathbf{X}' - \mathbf{p}\mathbf{f}_{\mathbf{x}}(\mathbf{x}, \mathbf{u})\mathbf{X}]\,dt.$$

Thanks to the system that $\mathbf{X}$ satisfies, we can further put

$$\int_0^T F_{\mathbf{x}}(\mathbf{x}, \mathbf{u})\mathbf{X}\,dt = \int_0^T [\mathbf{p}\mathbf{f}_{\mathbf{x}}(\mathbf{x}, \mathbf{u})\mathbf{X} + \mathbf{p}\mathbf{f}_{\mathbf{u}}(\mathbf{x}, \mathbf{u})\mathbf{U} - \mathbf{p}\mathbf{f}_{\mathbf{x}}(\mathbf{x}, \mathbf{u})\mathbf{X}]\,dt,$$

that is to say

$$\int_0^T F_{\mathbf{x}}(\mathbf{x}, \mathbf{u})\mathbf{X}\,dt = \int_0^T \mathbf{p}\mathbf{f}_{\mathbf{u}}(\mathbf{x}, \mathbf{u})\mathbf{U}\,dt.$$

Hence (7.1) becomes

$$\int_0^T [F_{\mathbf{u}}(\mathbf{x}, \mathbf{u}) + \mathbf{p}\mathbf{f}_{\mathbf{u}}(\mathbf{x}, \mathbf{u})]\mathbf{U}\,dt.$$

If $\mathbf{u}$ is indeed an optimal solution, then this variation on the cost produced by an arbitrary perturbation $\mathbf{U}$ must vanish, and this in turn leads to the condition

$$F_{\mathbf{u}}(\mathbf{x}, \mathbf{u}) + \mathbf{p}\mathbf{f}_{\mathbf{u}}(\mathbf{x}, \mathbf{u}) \equiv \mathbf{0}$$

due to such arbitrariness of $\mathbf{U}$. We have justified the following statement.

Proposition 7.1 *Assume* $\mathbf{u}$ *is an optimal solution of the control problem stated at the beginning of this section. Then*

$$F_{\mathbf{u}}(\mathbf{x}, \mathbf{u}) + \mathbf{p}\mathbf{f}_{\mathbf{u}}(\mathbf{x}, \mathbf{u}) \equiv \mathbf{0}$$

if

$$-\mathbf{p}'(t) = \mathbf{p}(t)\mathbf{f}_{\mathbf{x}}(\mathbf{x}(t), \mathbf{u}(t)) + F_{\mathbf{x}}(\mathbf{x}(t), \mathbf{u}(t)) in (0, T), \quad \mathbf{p}(T) = \mathbf{0},$$
$$\mathbf{x}'(t) = \mathbf{f}(\mathbf{x}(t), \mathbf{u}(t)) in (0, T), \quad \mathbf{x}(0) = \mathbf{x}_0.$$

This main conclusion is traditionally stated in terms of the so-called hamiltonian of the problem. If we set

$$H(\mathbf{x}, \mathbf{u}, \mathbf{p}) = F(\mathbf{x}, \mathbf{u}) + \mathbf{p} \cdot \mathbf{f}(\mathbf{x}, \mathbf{u}),$$

then, an optimal pair $(\mathbf{x}, \mathbf{u})$ goes with an optimal costate $\mathbf{p}$ so that

$$\frac{\partial H}{\partial \mathbf{u}} = \mathbf{0}, \quad \mathbf{p}' = -\frac{\partial H}{\partial \mathbf{x}}, \quad \mathbf{x}' = \frac{\partial H}{\partial \mathbf{p}},$$

together with $\mathbf{x}(0) = \mathbf{x}_0$, $\mathbf{p}(T) = \mathbf{0}$. This last terminal condition for the costate $\mathbf{p}$ is referred to as the transversality condition. The reason for this terminology (hamiltonian) is clear. The algebraic condition implies that indeed the hamiltonian is independent of $\mathbf{u}$ along optimal solutions, and the other two sets of differential equations involved have hamiltonian structure. Therefore the hamiltonian itself, evaluated on such an optimal solution $(\mathbf{x}(t), \mathbf{u}(t), \mathbf{p}(t))$ is constant in time because

$$\frac{d}{dt} H(\mathbf{x}(t), \mathbf{u}(t), \mathbf{p}(t)) = \frac{\partial H}{\partial \mathbf{x}} \cdot \mathbf{x}' + \frac{\partial H}{\partial \mathbf{p}} \cdot \mathbf{p}' \equiv 0.$$

As a matter of fact, if the terminal time T is not determined, then one has to impose the additional condition

$$H((\mathbf{x}(t), \mathbf{u}(t), \mathbf{p}(t)) \equiv 0$$

for all times t, in particular for $t = 0$, and $t = T$.

If a terminal constraint of the type $\mathbf{x}(T) = \mathbf{x}_T$ is to be enforced, then this condition replaces the transversality condition $\mathbf{p}(T) = \mathbf{0}$. Suppose that the problem to be solved is to

$$\text{Minimize in } \mathbf{u}(t) : \quad I(\mathbf{u}) = \int_0^T F(\mathbf{x}(t), \mathbf{u}(t)) \, dt$$

under

$$\mathbf{x}'(t) = \mathbf{f}(\mathbf{x}(t), \mathbf{u}(t)) \text{ in } (0, T), \quad \mathbf{x}(0) = \mathbf{x}_0, \mathbf{x}(T) = \mathbf{x}_T.$$

We have an improved version of the last proposition.

Proposition 7.2 *Assume* $\mathbf{u}$ *is an optimal solution of this control problem under a terminal time condition. Then*

$$F_{\mathbf{u}}(\mathbf{x}, \mathbf{u}) + \mathbf{p}\mathbf{f}_{\mathbf{u}}(\mathbf{x}, \mathbf{u}) \equiv \mathbf{0},$$
$$-\mathbf{p}'(t) = \mathbf{p}(t)\mathbf{f}_{\mathbf{x}}(\mathbf{x}(t), \mathbf{u}(t)) + F_{\mathbf{x}}(\mathbf{x}(t), \mathbf{u}(t)) \text{ in } (0, T),$$
$$\mathbf{x}'(t) = \mathbf{f}(\mathbf{x}(t), \mathbf{u}(t)) \text{ in } (0, T), \quad \mathbf{x}(0) = \mathbf{x}_0, \mathbf{x}(T) = \mathbf{x}_T.$$

If T is to be determined optimally, then

$$H((\mathbf{x}(t), \mathbf{u}(t), \mathbf{p}(t)) \equiv 0$$

for all times.

Example 7.4 Consider the simple, specific situation

$$\text{Minimize in } u(t) \in \mathbb{R} : \quad \int_0^1 u(t)^2 \, dt$$

under

$$x'(t) = x(t) + u(t) \text{ in } (0, 1), \quad x(0) = 1, x(1) = 0.$$

The hamiltonian of the problem is

$$H(x, u, p) = u^2 + p(x + u),$$

and optimality conditions amount to

$$2u + p = 0, \quad p' = -p, \quad x' = x + u,$$

together with $x(0) = 1$, $x(1) = 0$. It is immediate to have

$$u = -p/2, \quad p = -2ce^{-t}, \quad x' = x + ce^{-t}.$$

We have written $-2c$ for the arbitrary constant of the general solution of $p' = -p$, to avoid denominators and minus signs. We know from the basic theory for Linear Differential Equations of constant coefficients that the general solution of the equation for x is

$$x(t) = Ce^t - \frac{c}{2}e^{-t}$$

for constants C, c to be determined to adjust initial and final conditions. It is easy to find that

$$C = \frac{1}{1 - e^2}, \quad c = \frac{2e^2}{1 - e^2},$$

and then the optimal strategy and its corresponding optimal state are

$$u(t) = \frac{2}{1 - e^2}e^{2-t}, \quad x(t) = \frac{1}{1 - e^2}(e^t - e^{2-t}).$$

Example 7.5 This is taken from [10]. Gout is produced by excess of uric acid in blood. Suitable drugs are to be used to reduce its presence to acceptable levels. If $x(t)$ stands for the content of this acid in blood (in appropriate units), suppose we find at a certain time $t = 0$ that $x(0) = 1$. Being this level unacceptable, we set to ourselves the goal of bringing this level to zero $x(T) = 0$ in a future time T through the use $u(t)$ of the drug administered at time t. Under these circumstances $x(t)$ evolves according to the law $x(t)' = -x(t) + 1 - u(t)$. Bearing in mind that an excessive dosage of the drug may have undesirable side effects, and may be too expensive, we propose an objective functional that looks for a balance between the dosage of the drug, and the time T when we bring $x(T) = 0$, through the expression

$$\int_0^T \frac{1}{2}(k^2 + u(t)^2)\, dt.$$

k is a positive constant that measures the relative importance between those two objectives.

The hamiltonian of the problem is

$$H(x, u, p) = \frac{1}{2}(k^2 + u^2) + p(-x + 1 - u).$$

Optimality conditions read

$$u - p = 0, \quad p' = p, \quad x' = -x + 1 - u,$$

and we have $x(0) = 1, x(T) = 0$, but T is not known, and so it is part of the optimal solution. Computations can be easily made explicit. We have

$$p = u = ce^t, \quad x' = -x + 1 - ce^t.$$

Because T is not known, we need to enforce the condition

$$0 = H(x(0), u(0), p(0)) = \frac{1}{2}(k^2 + c^2) + c(-1 + 1 - c) = \frac{1}{2}(k^2 - c^2).$$

We conclude that $c = k$ (why does the possibility $c = -k$ have to be discarded?). The general solution of the equation $x' = -x + 1 - ke^t$ is $x(t) = 1 - (k/2)e^t + Ce^{-t}$. The initial condition $x(0) = 1$ leads to the value $C = k/2$, and so $x(t) = 1 + (k/2)(e^{-t} - e^t)$. T is the value at which this function vanishes $x(T) = 0$ yielding

$$T = \ln\left(\frac{1}{k} + \sqrt{1 + \frac{1}{k^2}}\right).$$

This is the time needed to eliminate completely uric acid in blood, but only if the treatment is given by the optimal dosage $u(t) = ke^t$ until $t = T$. If not, then a longer period would be necessary. One clearly sees that as k becomes larger and larger, giving priority to accomplishing $x(T) = 0$ regardless of price and other considerations, T becomes smaller and smaller. But as k becomes smaller and smaller, T grows indefinitely.

7.4 Pontryaguin's Principle

We next focus on a typical optimal control problem where a given compact and convex set $\mathbf{K}$ is to be taken into account. The model problem is

$$\text{Minimize in } \mathbf{u}(t) \in \mathbf{K}: \quad I(\mathbf{u}) = \int_0^T F(\mathbf{x}(t), \mathbf{u}(t))\, dt$$

subject to

$$\mathbf{x}(t)' = \mathbf{f}(\mathbf{x}(t), \mathbf{u}(t)) \text{ in } (0, T), \quad \mathbf{x}(0) = \mathbf{x}_0, (\mathbf{x}(T) = \mathbf{x}_T).$$

The fundamental result is Pontryaguin's principle. It is also stated in terms of the the hamiltonian

$$H(\mathbf{x}, \mathbf{u}, \mathbf{p}) = F(\mathbf{x}, \mathbf{u}) + \mathbf{p}\mathbf{f}(\mathbf{x}, \mathbf{u}).$$

Theorem 7.2 *Let* $\mathbf{u}$ *be an optimal solution of this problem, with associated (optimal) state* $\mathbf{x}$. *Then there is* $\mathbf{p}$, *the optimal costate, so that the triplet* $(\mathbf{u}(t), \mathbf{x}(t), \mathbf{p}(t))$ *is a solution of*

$$\mathbf{p}'(t) = -\frac{\partial H}{\partial \mathbf{x}}(\mathbf{x}(t), \mathbf{u}(t), \mathbf{p}(t)), \quad \mathbf{x}'(t) = \frac{\partial H}{\partial \mathbf{p}}(\mathbf{x}(t), \mathbf{u}(t), \mathbf{p}(t)),$$

$$H(\mathbf{x}(t), \mathbf{u}(t), \mathbf{p}(t)) = \min_{\mathbf{v} \in \mathbf{K}} H(\mathbf{x}(t), \mathbf{v}, \mathbf{p}(t))$$

in $(0, T)$, *together with* $\mathbf{x}(0) = \mathbf{x}_0$, *and one of the two conditions* $\mathbf{x}(T) = \mathbf{x}_T$, $\mathbf{p}(T) = \mathbf{0}$, *for each component, at the terminal time* T.

We notice that the condition on the minimum of the hamiltonian replaces the algebraic equation $\partial H/\partial \mathbf{u} = 0$. They both are coherent because the minimum of a convex function is always attained where the derivative vanishes.

What we are interested in is to discuss how the condition on the minimum is utilized and manipulated in practice.

Example 7.6 Let us start with a typical positioning problem for a double integrator. The task consists in traveling a given distance $L > 0$, and stopping, starting from rest, in a time as short as possible, assuming that we have a maximum rate $b > 0$

of acceleration, and a maximum braking rate $-a$, $a > 0$. The control $u(t)$ is the accelerator $x''(t) = u(t)$, and we seek to have $x(T) = L$, $x'(T) = 0$, having $x(0) = x'(0) = 0$, for $T > 0$ as small as possible. The function $x(t)$ indicates the position of the mobile with respect to the position at time $t = 0$.

It is standard to reduce the second-order law $x'' = u$ to a first-order system in order to write the hamiltonian. Indeed, the problem is to

$$\text{Minimize in } u(t) \in [-a, b] : \quad \int_0^T 1 \, dt$$

subject to

$$x_1'(t) = x_2(t), x_2'(t) = u(t) \text{ in } (0, T), \quad x_1(0) = x_2(0) = 0, x_1(T) = L, x_2(T) = 0.$$

T is unknown, and, indeed, an important piece of information for the optimal solution. The hamiltonian is therefore written as

$$H(x_1, x_2, u, p_1, p_2) = 1 + p_1 x_2 + p_2 u,$$

noticing that, because the state law is a first-order system of two ODEs, we will have two components for the state $\mathbf{x} = (x_1, x_2)$, and two corresponding components $\mathbf{p} = (p_1, p_2)$ for the costate. Once we have the hamiltonian, we proceed in three steps to examine Pontryaguin's principle.

1. We first look at the condition for the minimum of the hamiltonian with respect to the control variable u. According to the statement of Theorem 7.2, we should care about the minimum of H with respect to u over the interval $[-a, b]$. The important observation is that, since H is affine on u, the minimum will be necessarily attained either at the left-end point $-a$ whenever the coefficient $p_2(t)$ is positive; or at the right-end point b if $p_2(t)$ is negative. If it turns out to be zero, then, in principle, we can select u arbitrarily in $[-a, b]$. Hence, the optimal strategy is

$$u(t) = \begin{cases} -a, & \text{if } p_2(t) > 0, \\ b, & \text{if } p_2(t) < 0. \end{cases}$$

We therefore conclude that:

- The optimal strategy $u(t)$ can only make use of the two end-points of the interval $[-a, b]$, and to express this situation one talks about an optimal strategy of type bang-bang. This is always the case whenever the dependence of the hamiltonian H on the control variable u is linear.
- Changes from one end-point to the other take place when the costate p_2 changes sign, basically, when it goes through zero. The function, in this example $p_2(t)$, indicating when and how the optimal strategy jumps from one end-point to the other is called the switching function of the problem.

2. To move on, we need information about $p_2(t)$, and, in particular about when it can change sign. To this end, we examine the ODEs that costates must verify

$$p_1'(t) = 0, \ p_2'(t) = -p_1(t) \text{ in } (0, T).$$

There is no transversality condition because we are asking for terminal conditions as well. We conclude that p_1 is a constant, and then $p_2(t) = c - p_1 t$. Though we do not have information about those two constants c, and p_1, yet the form of $p_2(t)$ as a linear function of t suffices to conclude that there is, at most, one change in sign for p_2, and, therefore, at most one change for the optimal $u(t)$ from $-a$ to b, or from b to $-a$. The interpretation of the problem very clearly suggests that we need to start with $u = b$, so that the mobile moves ahead, and at a certain time t_0, change to $u = -a$ and be sure that it will stop when reaching the point $x = L$. Hence, we conclude

$$u(t) = \begin{cases} b, & \text{if } 0 < t < t_0, \\ -a, & \text{if } t_0 < t < T. \end{cases}$$

Notice the difference between the form of the optimal $u(t)$ from the first step to the second.

3. We finally examine the state law $x''(t) = u(t)$ for u of the form at the end of the previous step. Because $u(t)$ has two constant values in different subintervals, we need to proceed in two substeps.

   (a) Interval $[0, t_0]$. We start from rest $x(0) = x'(0) = 0$, and $x'' = b$ all over this subinterval. By integrating twice, and using the starting conditions, we immediately find that $x(t) = (b/2)t^2$. At the end of this subinterval the state of our mobile object is

   $$x(t_0) = \frac{b}{2}t_0^2, \quad x'(t_0) = bt_0.$$

   (b) Interval $[t_0, T]$. We retake, as initial conditions for this second period, the conditions at t_0 that we left at the end of the previous one, and solve now the problem

   $$x''(t) = -a \text{ in } (t_0, T), \quad x(t_0) = \frac{b}{2}t_0^2, x'(t_0) = bt_0.$$

   The solution is

   $$x(t) = -\frac{a}{2}(t - t_0)^2 + bt_0(t - t_0) + \frac{b}{2}t_0^2.$$

   But this function must be such that terminal time conditions do hold $x(T) = L, x'(T) = 0$. After going through the algebra, one finds that

$$t_0 = \sqrt{\frac{2aL}{b(a+b)}}, \quad T = \sqrt{\frac{2(a+b)L}{ab}}.$$

We conclude that the shortest time in which the task can be performed is this value of T, but only if the accelerator is used according to the optimal $u(t)$. Different uses will employ a longer time.

Example 7.7 We look again at a shortest time problem. This time we have the law for a harmonic oscillator

$$x''(t) + x(t) = u(t), \quad u(t) \in [-1, 1].$$

We would like to take, as soon as possible, the state (x, x') from an arbitrary initial state (x_0, x_0') to rest $(0, 0)$. If we put $\mathbf{x} = (x_1, x_2) = (x, x')$, the second-order state equation becomes the first-order system

$$x_1' = x_2, \quad x_2' = -x_1 + u,$$

and the hamiltonian is

$$H(x_1, x_2, u, p_1, p_2) = 1 + p_1 x_2 + p_2(-x_1 + u).$$

As in the example above, the linear dependence of H with respect to u, implies that the optimal control $u(t)$ is necessarily of bang-bang type, jumping between the two extremes -1 and 1. Since the coefficient multiplying u is the costate p_2, once again the changes from 1 to -1, or from -1 to 1, are dictated by every time that p_2 passes through zero. What is quite different, with respect to the double integrator, is the set of differential equations that costates have to verify. Namely,

$$p_1' = p_2, \; p_2' = -p_1 \text{ in } (0, T).$$

If we eliminate p_1, since we are interested in p_2, we have that $p_2'' + p_2 = 0$, and we are back to the homogeneous version of the state equation. It is well-known that the general solution of this equation is a linear combination of $\cos t$, and $\sin t$. Equivalently, we can also write $p_2(t) = A \cos(t + B)$ for A, B, arbitrary constants. This is a big difference with the previous example. Two main conclusions are:

- p_2 can change sign an arbitrary number of times, possibly depending on the initial condition (x_0, x_0');
- p_2 cannot keep the same sign longer than π units.

An additional, important piece of information is the set of integral curves of both extremal possibilities

$$x'' + x = 1, \quad x'' + x = -1.$$

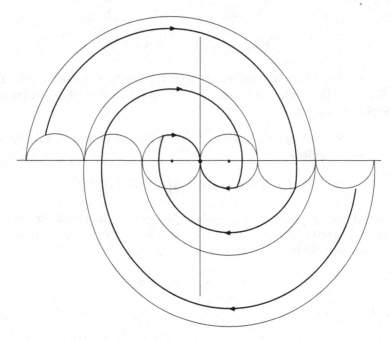

Fig. 7.1 Optimal strategies for the harmonic oscillator

It is easy to argue that the first one corresponds to a family of concentric circles around $(1, 0)$ run clockwise, while the second is a similar family of circles around $(-1, 0)$. After all, as before, the general solution of the homogeneous second-order equation $x'' + x = 0$ can be written in the form

$$x(t) = r \cos(t + \theta), \quad x'(t) = -r \sin(t + \theta),$$

with $r > 0$ the radius of the circle, and θ, the phase. The issue is then to understand how these two families of circles are to be mixed so as to bring any point in the plane to the origin by jumping from circles of one family to circles of the other, knowing that, at most, we can stay in half of each circle. It is a matter of a little bit of reflection to realize how this can be accomplished. Figure 7.1 may help in visualizing the optimal strategies. This is a classical picture that can be seen in almost every introductory text to optimal control.

Example 7.8 A rocket is launched with the aim of hitting a target located $3 + 5/6$ units away in a straight line in 3 units of time. It must be accomplished with a least amount of fuel measured by the integral

$$\int_0^3 k u(t)^2 \, dt$$

where $u(t) \in [0, 1]$ is the way we use the accelerator

$$x''(t) = u(t) \quad t \in (0, 3),$$

and $k > 0$ is a constant. As in a previous example, reducing the scalar second-order state law to a first-order system through the variables

$$x_1 = x, \quad x_2 = x',$$

we can formulate the problem as to

$$\text{Minimize in } u(t) \in [0, 1]: \quad k \int_0^3 u(t)^2 \, dt$$

under

$$x_1'(t) = x_2(t), x_2'(t) = u(t) \text{ in } (0, 3),$$
$$x_1(0) = x_2(0) = 0, x_1(3) = 3 + 5/6.$$

This time $x_2(3)$ is not predetermined as it is a hitting problem, not a soft adjustment. The hamiltonian of the problem is

$$H(x_1, x_2, u, p_1, p_2) = ku^2 + p_1 x_2 + p_2 u.$$

The new feature of this situation is the quadratic dependence of H on u. We proceed as in a previous example in three succesive steps.

1. We first look at the minimum of H with respect to $u \in [0, 1]$ to find that

$$u(t) = \begin{cases} 0, & \text{if } -p_2(t)/(2k) \leq 0, \\ -p_2(t)/(2k), & \text{if } -p_2(t)/(2k) \in [0, 1], \\ 1, & \text{if } -p_2(t)/(2k) \geq 1, \end{cases}$$

where $u = -p_2(t)/(2k)$ is precisely the vertex (the minimum point) of the hamiltonian. To proceed, we need more information about the multiplier $p_2(t)$.

2. Differential equations for the costates

$$p_1'(t) = 0, p_2'(t) = -p_1(t) \text{ in } (0, 3).$$

We conclude that p_1 is a constant, and then $p_2(t) = -p_1 t + c$ for some constant c. This time however, we need to impose a transversality condition for p_2 at $t = 3$ because $x_2(3)$ is free. We must have $p_2(3) = 0$, and so $p_2(t) = d(3 - t)$, if we put $d = p_1$, a constant. Once we have this information on $p_2(t)$, we see that it can, at most, go through 0 and 1 just once. If we incorporate the constant $-1/(2k)$ into the constant d, we would have

$$u(t) = \begin{cases} 0, & \text{if } d(3-t) \leq 0, \\ d(3-t), & \text{if } d(3-t) \in [0, 1], \\ 1, & \text{if } d(3-t) \geq 1, \end{cases}$$

If the constant d were non-positive, then, since $3 - t \geq 0$ over the interval $[0, 3]$, we would have $u \equiv 0$, and the rocket would not move. We ought to have then $d > 0$, and

$$u(t) = \begin{cases} 1, & \text{if } 0 < t < t_0, \\ d(3-t), & \text{if } t_0 < t < 3, \end{cases}$$

for a certain time t_0 where $d(3 - t_0) = 1$.

3. Based on this information for the optimal control $u(t)$, we now turn to solving the state equation to determine the instant t_0, and whatever additional information that may be relevant. We need to proceed in two subintervals.

   (a) In the initial subinterval $[0, t_0]$, we have $x''(t) = 1$ with $x(0) = x'(0) = 0$. The solution is $x(t) = t^2/2$, leaving the rocket at the state $x(t_0) = t_0^2/2$, $x'(t_0) = t_0$.

   (b) In the second subinterval $[t_0, 3]$, bearing in mind that $d(3 - t_0) = 1$, we would have $x''(t) = d(3 - t)$ with $x(t_0) = t_0^2/2$, $x'(t_0) = t_0$. With a bit of care in the calculations, and recalling that $d(3 - t_0) = 1$, we find that

$$x(t) = -\frac{d}{6}(t - t_0)^3 + \frac{1}{2}(t - t_0)^2 + t_0 t - \frac{1}{2}t_0^2.$$

By demanding $3 + 5/6 = x(3)$, we find $d = 1/2$, and $t_0 = 1$.

The optimal strategy is then

$$u(t) = \begin{cases} 1, & \text{if } 0 < t < 1, \\ (3-t)/2, & \text{if } 1 < t < 3. \end{cases}$$

7.5 Other Important Constraints

The most general optimal control problem we will treat in this text is

$$\text{Minimize in } \mathbf{u}(t) \in \mathbf{K}: \quad I(\mathbf{u}) = \int_0^T F(\mathbf{x}(t), \mathbf{u}(t))\, dt + \phi(\mathbf{x}(T))$$

under

$$\mathbf{x}(t)' = \mathbf{f}(\mathbf{x}(t), \mathbf{u}(t)) \text{ in } (0, T), \quad \mathbf{x}(0) = \mathbf{x}_0.$$

How does this new ingredient, the presence in the cost of a term depending on the terminal state, affect Pontryaguin's maximum principle?

It is not hard to understand a short discussion about this issue. Notice that the Fundamental Theorem of Calculus implies

$$\phi(\mathbf{x}_T) - \phi(\mathbf{x}_0) = \int_0^T \frac{d}{dt}\phi(\mathbf{x}(t)) \, dt.$$

But the right-hand side is, by the chain rule and the state law,

$$\int_0^T \nabla\phi(\mathbf{x}(t)) \cdot \mathbf{x}'(t) \, dt = \int_0^T \nabla\phi(\mathbf{x}(t)) \cdot \mathbf{f}(\mathbf{x}(t), \mathbf{u}(t)) \, dt.$$

Because the constant term $\phi(\mathbf{x}_0)$ does not affect minimization, we see that our new problem can be recast in the form

$$\text{Minimize in } \mathbf{u}(t): \quad \tilde{I}(\mathbf{u}) = \int_0^T \tilde{F}(\mathbf{x}(t), \mathbf{u}(t)) \, dt$$

under

$$\mathbf{x}'(t) = \mathbf{f}(\mathbf{x}(t), \mathbf{u}(t)) \text{ in } (0, T), \quad \mathbf{x}(0) = \mathbf{x}_0,$$

for

$$\tilde{F}(\mathbf{x}, \mathbf{u}) = F(\mathbf{x}, \mathbf{u}) + \nabla\phi(\mathbf{x}) \cdot \mathbf{f}(\mathbf{x}, \mathbf{u}).$$

Curiously enough, this function $\tilde{F}$ has a structure similar to the hamiltonian of the problem. By Pontryaguin's maximum principle, applied to $\tilde{F}$, to every optimal pair $(\mathbf{x}, \mathbf{u})$ we can associate an optimal costate $\mathbf{p}$ such that optimality conditions in Theorem 7.2 hold for the hamiltonian

$$\tilde{H}(\mathbf{x}, \mathbf{u}, \mathbf{p}) = F(\mathbf{x}, \mathbf{u}) + \nabla\phi(\mathbf{x}) \cdot \mathbf{f}(\mathbf{x}, \mathbf{u}) + \mathbf{p} \cdot \mathbf{f}(\mathbf{x}, \mathbf{u}).$$

Since we can write

$$\tilde{H}(\mathbf{x}, \mathbf{u}, \mathbf{p}) = F(\mathbf{x}, \mathbf{u}) + (\nabla\phi(\mathbf{x}) + \mathbf{p}) \cdot \mathbf{f}(\mathbf{x}, \mathbf{u}),$$

we realize that in fact the optimal costate $\mathbf{p}(t)$ for the original problem can be understood through the optimal costate $\tilde{\mathbf{p}} = \nabla\phi(\mathbf{x}) + \mathbf{p}$ for the same optimal control problem without the term $\phi(\mathbf{x}(T))$ in the cost, but enforcing the transversality condition

$$\tilde{\mathbf{p}}(T) = \nabla\phi(\mathbf{x}(T)).$$

Corollary 7.1 *Let* **u** *be an optimal solution of the problem with the term* $\phi(\mathbf{x}(T))$ *in the cost. If* $(\mathbf{u}, \mathbf{x}, \mathbf{p})$ *is the optimal triplet, then*

$$\mathbf{p}'(t) = -\frac{\partial H}{\partial \mathbf{x}}(\mathbf{x}(t), \mathbf{u}(t), \mathbf{p}(t)), \quad \mathbf{x}'(t) = \frac{\partial H}{\partial \mathbf{p}}(\mathbf{x}(t), \mathbf{u}(t), \mathbf{p}(t)),$$

$$H(\mathbf{x}(t), \mathbf{u}(t), \mathbf{p}(t)) = \min_{\mathbf{v} \in \mathbf{K}} H(\mathbf{x}(t), \mathbf{v}, \mathbf{p}(t))$$

in $(0, T)$*, together with* $\mathbf{x}(0) = \mathbf{x}_0$*, and* $\mathbf{p}(T) = \nabla\phi(\mathbf{x}(T))$ *at the terminal time* T*.*

We therefore see that the optimal solution is sought exactly in the same way as if the contribution $\phi(\mathbf{x}(T))$ were not present. This term only enters into the transversality condition that is changed from $\mathbf{p}(T) = \mathbf{0}$ to $\mathbf{p}(T) = \nabla\phi(\mathbf{x}(T))$.

We now have the tools to find the optimal solution of the farmer and the honey situation at the beginning of the chapter.

Example 7.9 In order to write the problem as a minimization situation instead of maximizing net income, we change the sign of the net income, and seek to

$$\text{Minimize in } u(t) \in [0, 1]: \quad \int_0^1 \frac{1}{2}u(t)^2 \, dt - \frac{2}{e}x(1)$$

subject to

$$x'(t) = x(t) + u(t) \text{ in } (0, 1), \quad x(0) = 1/4.$$

The hamiltonian is

$$H(x, u, p) = \frac{1}{2}u^2 + p(x + u)$$

which is quadratic in u. The point of minimum is $u = -p$, and so, for the minimum of H over the interval $u \in [0, 1]$, we can put

$$u(t) = \begin{cases} 0, & \text{if } -p \le 0, \\ -p, & \text{if } -p \in [0, 1], \\ 1, & \text{if } -p \ge 1, \end{cases} = \begin{cases} 0, & \text{if } p \ge 0, \\ -p, & \text{if } -1 \le p \le 0, \\ 1, & \text{if } p \le -1. \end{cases}$$

We next examine the differential equation that p should comply with $p'(t) = -p(t)$, with the new transversality condition $p(1) = -2/e$. Note that in this example

$$\phi(x) = -\frac{2}{e}x, \quad \phi'(x) = -\frac{2}{e}.$$

It is interesting to notice that this time we can fully determine the costate $p(t) = -2e^{-t}$, thanks to this transversality condition. Upon examining the properties of this function, in particular the range where it lies over $(-\infty, -1]$, $[-1, 0]$, and $[0, \infty)$, we are sure that

$$u(t) = \begin{cases} 1, & \text{if } 0 < t < \ln 2, \\ 2e^{-t}, & \text{if } \ln 2 < t < 1. \end{cases}$$

In particular, the optimal strategy is not to introduce outer individuals at maximum rate into the hives. Can our readers find the improvement on the net income for the optimal control compared to the possibility $u(t) \equiv 1$ all over $[0, 1]$?

7.6 Sufficiency of Optimality Conditions

We have stressed in all of the examples worked out in the preceding sections how the use of optimality conditions can help in finding the true optimal strategies, at least, for well-prepared situations. In all cases examined, there was a unique optimal solution and it was taken for granted that that was the optimal solution sought. As a matter of fact, if we trust Theorem 7.1, under those hypotheses on the various ingredients of an optimal control problem, there are optimal solutions. If on the other hand, there is one unique solution of optimality conditions, then that one has to be the optimal solution. In a more precise and explicit way, it is relevant to pay some attention to assumptions ensuring that solutions of optimality conditions are indeed optimal. As in previous chapters, this is the important issue of sufficiency of optimality conditions.

Theorem 7.3 *Under the same hypotheses as in Theorem 7.1, namely*

1. the cost integrand $F(t, \mathbf{x}, \mathbf{u})$ is a convex function regarded as a function of pairs $(\mathbf{x}, \mathbf{u})$ for every time t, and coercive in the sense

$$\lim_{|\mathbf{u}| \to \infty} \frac{F(t, \mathbf{x}, \mathbf{u})}{|\mathbf{u}|} = +\infty,$$

uniformly with respect to every time t, and every vector $\mathbf{x}$;
2. the mapping $\mathbf{f}(t, \mathbf{x}, \mathbf{u})$ is a linear mapping regarded as a mapping of pairs $(\mathbf{x}, \mathbf{u})$ for every time t;
3. the set $\mathbf{K}$ is closed and convex;

every solution of the corresponding optimality conditions is optimal for the control problem. If, in addition, the dependence of $F(\mathbf{x}, \mathbf{u})$ on the control variable $\mathbf{u}$ is strictly convex, then there cannot be more than one optimal solution.

The proof of this result is deferred to a more advanced course.

7.7   Final Remarks

It is not difficult to imagine that optimal control is an area of considerable interest in applications. Even though we have treated very specific and academic situations, one can see the wide variety of situations one can consider. It is not an easy subject because pretty soon computations become impossible by hand, and approximation techniques of different nature need to take over. This will be the focus of our last chapter. In realistic problems, difficult ingredients that we have avoided in this first exposure are fundamental: non-linearities of various kinds, constraints in the state variables, more general cost functionals, the need to react instantaneously to unexpected disturbances, the occurrence of singular controls, etc. But all of this is part of deeper treatises on the subject. See some additional references in the bibliography section.

7.8   Exercises

7.8.1   Exercises to Support the Main Concepts

1. Consider the optimal control problem

$$\text{Minimize in } u(t): \quad \frac{1}{2} \int_0^1 u(t)^2 \, dt$$

   subjected to

$$x'(t) = x(t) + u(t) \text{ in } (0, 1), \quad x(0) = 1, x(1) = 0.$$

   Find the solution by eliminating u from the formulation, and seeking the optimal $x(t)$ through the resulting variational problem.
2. Suppose the ingredients of a given optimal control problem are such that the corresponding hamiltonian

$$H(u, \mathbf{x}, \mathbf{p}) = F(\mathbf{x}, u) + \mathbf{p}\mathbf{f}(\mathbf{x}, u)$$

   turns out to be linear in the control variable, i.e.

$$F(\mathbf{x}, u) = F_0(\mathbf{x}) + F_1(\mathbf{x})u, \quad \mathbf{f}(\mathbf{x}, u) = \mathbf{f}_0(\mathbf{x}) + \mathbf{f}_1(\mathbf{x})u.$$

   The coefficient in front of u in the hamiltonian is called the switching function. Show that the sign of this function determines the value of the optimal $u(t)$ under a constraint $u(t) \in [\alpha, \beta]$.

3. A control system governed by the linear law

$$\mathbf{x}'(t) = \mathbf{A}\mathbf{x}(t) + \mathbf{U}$$

can only accept a constant control vector $\mathbf{U}$. Find the best such possibility to take the system from an initial state $\mathbf{x}_0$ to a final, arbitrary state $\mathbf{x}_T$, in a time horizon T. Apply your conclusion to a one-dimensional situation $\mathbf{x} = x, \mathbf{U} = U, \mathbf{A} = a$.

4. Retake the problem of the double integrator $x''(t) = u(t) \in [-a, b]$ with initial condition $x(0) = x'(0) = 0$, and desired final state $x(T) = L, x'(0) = 0$. Given that we know that the optimal strategy is of bang-bang type $u(t) = b$ for $0 < t < t_0$, and $u(t) = -a$, for $t_0 < t < t_0 + t_1$, reinterpret the problem as minimizing the error

$$E(t_0, t_1) = \frac{1}{2}[x(t_0, t_1) - L]^2 + \frac{1}{2}[x'(t_0, t_1)]^2$$

in the pair of non-negative times (t_0, t_1). The state $(x(t_0, t_1), x'(t_0, t_1))$ is the corresponding final state for $t = t_0 + t_1$ for that control bang-bang.

5. Reconsider the first exercise

$$\int_0^1 \frac{1}{2}u^2 \, dt, \quad x' = x + u \text{ in } (0, 1),$$

and examine the connection between Pontryaguin's principle, and the E-L equation for the equivalent variational problem.

6. Consider the following problem

$$\text{Minimize in } u(t): \quad \int_0^T \frac{1}{2}u(t)^2 \, dt$$

subject to

$$x'(t) = -2x(t) + 10u(t) \text{ in } (0, T), \quad x(0) = x_0.$$

Can you figure out the optimal strategy $u(t)$ without any calculation? Think in terms of what is expected the system should perform. Compare it to the situation

$$\text{Minimize in } u(t): \quad \int_0^T \frac{1}{2}u(t)^2 \, dt$$

subject to

$$x'(t) = -2x(t) + 10u(t) \text{ in } (0, T), \quad x(0) = x_0, x(T) = x_T.$$

7. There is also a discrete version of the principle that leads to the Hamilton-Jacobi-Bellman equation. It is the basis of the field Dynamic Programming. We are given the following optimization problem

$$\text{Minimize in } (u_0, u_1): \quad x_0 u_0 + x_1 u_1 + x_2$$

where

$$x_{k+1} = x_k u_k + u_k, u_k \in [0, 1], k = 0, 1, \quad x_0 = 2.$$

(a) Write explicitly the cost functional in terms of (u_0, u_1), and find the optimal such point to minimize the cost.

(b) Write instead the cost functional in terms of (x_1, x_2) specifying the feasible region for these pairs. Find the optimal such pair, and check that it corresponds to the optimal (u_0, u_1).

7.8.2 Practice Exercises

1. Find the optimal strategy $u(t)$ to take a system described by $x(t)$ under the law $x''(t) + x(t) = u(t)$, from the initial condition $x(0) = 2, x'(0) = 0$ to rest $x(\pi) = x'(\pi) = 0$, by minimizing a quadratic cost

$$\frac{1}{2} \int_0^\pi u^2 \, dt.$$

2. Find the optimal control for the problem

$$\text{Minimize in } u(t): \quad \int_0^1 (x(t) + u(t)) \, dt$$

subject to

$$x'(t) = x(t) - u(t) \text{ in } (0, 1), \quad x(0) = 1, \quad |u(t)| \le 1.$$

3. A mobile object with position $x(t)$, measured with respect to its initial position $x(0) = 0$, starts out with velocity $x'(0) = 1$.

(a) The accelerator $u = x''$ is used in an optimal way so as to minimize the objective

$$\int_0^2 \left[x(t) + \frac{1}{2} u(t)^2 \right] dt,$$

and then the object stops at $T = 2$. What is the distance run?

(b) Consider the same problem under the constraint $|u(t)| \le 2$. Has the optimal strategy changed?

(c) Determine the smallest interval $[a, b]$ in such a way that the optimal strategy under the constraint $a \le u(t) \le b$ be the same as the first one computed above.

4. The initial position $x(0)$ of a vehicle with respect to its equilibrium position $x = 0$ is $x(0) = 1$, and its dynamics obeys the law $x'(t) = x(t) + u(t)$. We would like to determine the control law $u(t)$ capable of maintaining the position at 1 during the time interval $[0, 1]$.

   (a) Is it possible to find such a law under the constraint $u \in [-1/2, 1]$?
   (b) What is the optimal strategy to minimize the cost $(1/2)(x(1) - 1)^2$?
   (c) How do your answers change if we allow $u \in [-2, 2]$?

5. An aircraft is moving in a straight line away from its initial state $x(0) = 0$, $x'(0) = 1$, with an acceleration $u(t)$, $x''(t) = u(t)$. Find the optimal use $u(t)$ of the accelerator to minimize

$$\int_0^2 (1 + u(t)^2)\, dt$$

   in such a way that it stops at $T = 2$, $x'(2) = 0$. How far has the aircraft gone?

6. A given process evolves according to the state law

$$x'(t) = x(t) + u(t), \ t \in (0, 10),$$

   with initial state $x(0) = 100$. The cost is given by

$$I(x, u) = \frac{1}{2}x(10)^2 + \int_0^{10} (3x(t)^2 + u(t)^2)\, dt.$$

   (a) Suppose that we only allow for constant controls $u \equiv U$, independent of time. In this case the cost is quadratic in U, and can be expressed in the form $AU^2 + BU + C$. Calculate the values of the three coefficients A, B y C, but do not perform the resulting integrals.
   (b) Find the optimal control depending on time to minimize the cost.

7. Look at the optimal control problem

$$\text{Minimize in } u(t): \ \int_0^1 (x(t) + u^2(t))\, dt$$

   subject to

$$x'(t) = x(t) - u(t) \text{ in } (0, 1), \quad x(0) = 1, \quad |u(t)| \leq 1.$$

   (a) Compute the cost associated with the null control $u \equiv 0$.
   (b) Calculate the cost corresponding to the choice

$$u(t) = 1, \text{ for } 0 < t < 1/2; u(t) = -1, \text{ for } 1/2 < t < 1,$$

   and compare it to the latter to decide which one is better.
   (c) Find the best control and its cost.

8. A given process obeys the law $x' = -x + u$ where the control u is constrained by $|u| \leq 2$. An optimal use of the control u is desired in such a way that the area enclosed by $x(t)$ in $(0, 1)$ be minimal. Moreover, it is required that $x(0) = x(1) = 2$.

   (a) Formulate the situation as an optimal control problem identifying the various elements.
   (b) Is the control $u(t) \equiv 0$ admissible? Why?
   (c) Choose one feasible control $u(t)$ and calculate its cost.
   (d) Determine the optimal strategy and its cost.

7.8.3 Case Studies

1. The population $x(t)$ of a certain rare species would disappear if measures were not taken to preserve it. At the same time the current population $x(0) = P > 0$ is too large, and so harmful to its environment. The population can be controlled through $u(t)$, $-U \leq u(t) \leq U$ for some $U > 0$, where $x'(t) = -kx(t) + u(t)$, $k > 0$, a constant.

   (a) If a reduction to half its size $P/2$ is desired in the least time possible, determine the optimal strategy, and give the optimal time in terms of the constants of the problem P, U, k.
   (b) Under what conditions can a constant population $P/2$ can be maintained, and what strategy is to be followed?

2. An automatic control system for a harmful gas in a greenhouse is in charge of maintaining its state at the minimum $x = 0$ through the state law $x'(t) = \alpha x(t) + u(t)$ with $\alpha > 0$ where $x(t)$ stands for the density of such a gas, and $u(t)$ is the control acting on the system $-U \leq u \leq 0$, $U > 0$. Before an emergency $x(0) > 0$, the system reacts and tries to take back x to its vanishing value $x = 0$ in the least time possible.

   (a) Find the optimal strategy for $x(0) = C > 0$.
   (b) Determine the critical value C for which the system is incapable of taking back $x(0) = C$ to $x(T) = 0$ in a finite T.

3. A certain mechanism is vibrating. If $x(t)$ is the variable measuring such vibration with respect to its desired position at equilibrium, then

$$x''(t) + \omega_0^2 x(t) = \cos(\omega_0 t) + u(t).$$

If no measure is taken ($u \equiv 0$), the system would enter resonance from an initial state (x_0, x_0'). To avoid that undesirable phenomenon, a control u is sought minimizing its cost proportional to the integral of its square in a time interval $[0, \pi/w_0]$. Find the cheapest such way $u(t)$ to bring the system to rest, and avoid resonance.

4. A two-species, (x_1, x_2), habitat likes to stay at equilibrium which is the solution
   of the linear system
   $$\mathbf{A}\mathbf{x} + \mathbf{b} = \mathbf{0}, \quad \mathbf{x} = (x_1, x_2).$$

   Due to some disturbances, the system finds itself off-equilibrium
   $$\mathbf{A}\mathbf{x}(0) + \mathbf{b} \neq \mathbf{0}.$$

   Find the best way to take the system back to equilibrium in T units of time through
   the state system
   $$\mathbf{x}'(t) = \mathbf{A}\mathbf{x}(t) + \mathbf{b} + \mathbf{B}u(t),$$

   with a cost of the form
   $$\int_0^T \frac{1}{2}|u(t)|^2 \, dt + \frac{1}{2}|\mathbf{A}\mathbf{x}(T) + \mathbf{b}|^2.$$

   Is there any change in the solution if the control variable has two components
   $\mathbf{u} = (u_1, u_2)$ instead of one?
5. Gout is produced by an excess of uric acid in blood.[1] If $x(t)$ represents its
   concentration in the appropriate units, the law governing its change is $x'(t) =$
   $-x(t) + 1 - u(t)$ where $u(t)$ is the concentration of the treatment in charge of
   lowering the presence of uric acid. The level $x = 1$ is inadmissible. If for a certain patient we have $x(0) = 1$, determine the optimal dosage plan $u(t)$ to go from
   $x(0) = 1$ to $x(T) = 0$ in T units of time (known), if, to avoid side effects from a
   too aggressive a treatment, we demand $0 \leq u \leq 2$, and we minimize the price

   $$\frac{1}{2}\int_0^T u(t)^2 \, dt.$$

6. A police car is parked when a motorcycle passes by at a constant but excessive
   speed v. The car pursues the motorist trying to place itself parallel to the motorcycle and with its same velocity. This job is to be done in the least time possible,
   knowing that there is a maximum possible acceleration a, and maximum brake
   power $-b$.

   (a) Write in precise terms the optimal control problem.
   (b) Determine the hamiltonian.
   (c) Use Pontryaguin's principle to find the optimal strategy as far as it is possible
       (in terms of a system for two unknowns t_0, T).
   (d) Solve completely the problem in the particular case $a = b = v$.

---

[1] From [10].

7. A plague Y is destroying a crop X according to the law

$$x' = x - y, \quad y' = y$$

where $x(t)$ and $y(t)$ denote the density of population of X and Y for each time t, respectively. Initially, both are equal $x(0) = y(0) = 1$. To fight against the harmful plague, a predator U, with population $u(t)$, is introduced changing the dynamic laws to

$$x' = x - y - u/2, \quad y' = y - u, \quad 0 \le u \le 2.$$

(a) How long will the crop stay before vanishing if we do not act $u \equiv 0$?
(b) Argue that the life of X is longer for $u \equiv 1$.
(c) It is desirable to eliminate the population of Y in minimal time, but ensuring that X stays at half its initial size. Find the optimal stratey $u(t)$, and the time of extinction of the plague with that guarantee over the crop.

8. The position $x_1(t)$ off-equilibrium of a certain antenna can be controlled through the system[2]

$$\begin{pmatrix} x_1'(t) \\ x_2'(t) \end{pmatrix} = \begin{pmatrix} 0 & 1 \\ 0 & -1 \end{pmatrix} \begin{pmatrix} x_1(t) \\ x_2(t) \end{pmatrix} + \begin{pmatrix} 0 \\ 1 \end{pmatrix} u(t),$$

with no restriction on the size of the control u. If at some starting time we have $x_1(0) = 0.1, x_2(0) = -0.05$, we would like to take the antenna back to its equilibrium position $x_1 = x_2 = 0$ in a unit of time with a minimum expenditure measured by

$$\int_0^1 \frac{1}{2} u^2(t) \, dt.$$

9. Electric current in a certain circuit vanishes initially. A largest voltage difference

$$v(T) = \int_0^T R i'(t) \, dt$$

is desired where T is the given time horizon, and the dynamic law of the circuit is (R and L are constants)

$$i'(t) = \frac{u(t)}{L} - \frac{R}{L} i(t) \text{ in } (0, T), \quad u(t) \in [0, 1].$$

Determine the optimal use of the control variable u, and the largest possible voltage difference.

---

[2]From [5].

Chapter 8
Numerical Approximation of Basic Optimal Control Problems, and Dynamic Programming

8.1 Introduction

At this stage, and as it happened earlier in other chapters, it will not be a surprise to say that it is unavoidable to count on an efficient tool to approximate numerically the solution of optimal control problems. Except for very simple examples, where optimality conditions can be decoupled and solved one after the other, there is no hope that one can find optimal solutions to problems by naked hands. Yet, many of the usual techniques to design such numerical procedures rely on making good use of optimality conditions. These require to deal with costates and restrictions. From an approximation viewpoint is not that simple to manage constraints. They ask for quite a bit of expertise to produce accurate solutions. Our goal here, as in previous chapters, is much more modest. We just want to implement a simple tool that may be used to find reasonable approximations for typical examples.

The strategy is similar to the one describe for variational problems. As a matter of fact, we will deal with optimal control problems through their variational reformulation. A main ingredient of this format is the importance of enforcing point wise constraints, but this is precisely what we have been doing in the previous chapter. A very simple problem will help us understand what we mean by the previous sentences.

Assume we would like to solve the optimal control problem

$$\text{Minimize in } u(t) \in \mathbb{R}: \quad \int_0^1 u(t)^2 \, dt$$

under the state law $x'(t) = x(t) + u(t)$, and end-point conditions $x(0) = 1, x(1) = 0$. It is immediate to realize, by putting $u(t) = x'(t) - x(t)$, that this optimal control problem is equivalent to the variational problem

$$\text{Minimize in } x(t): \quad \int_0^1 [x'(t) - x(t)]^2 \, dt \quad \text{under} \quad x(0) = 1, x(1) = 0.$$

© Springer International Publishing AG 2017
P. Pedregal, *Optimization and Approximation*, UNITEXT 108,
DOI 10.1007/978-3-319-64843-9_8

We can now use either the corresponding Euler–Lagrange equation, or our basic approximation scheme for variational problems. Of course, the Euler–Lagrange equation for this problem is equivalent to the corresponding optimality conditions for the optimal control version. It is not difficult to conclude that the optimal control $u(t) = x'(t) - x(t)$ must be a solution of the differential equation $u(t)' + u(t) = 0$, and so $u(t) = ce^{-t}$ for some constant c. And then $x'(t) = x(t) + ce^{-t}$ with the two end-point conditions $x(0) = 1$, $x(1) = 0$. The general solution of this differential equation is

$$x(t) = de^t - \frac{c}{2}e^{-t}.$$

Adjusting the two end-point conditions lead to the optimal solution

$$x(t) = \frac{1}{e^2 - 1}(e^{2-t} - e^{-t}), \quad u(t) = \frac{-2e^2}{e^2 - 1}e^{-t}.$$

When there is an interval $K \subset \mathbb{R}$ to determine point wise constraints for feasible controls, we can still write down the equivalent variational problem, but the use of optimality conditions is not so straightforward. We can use our basic approximation techniques for variational problems under point wise constraints to find good approximations. This will be the way in which we will find reasonable numerical approximations for optimal control problems, in such a way that once a given optimal control situation of interest has been transformed into a variational problem under pointwise constraints, then one can apply the algorithm treated in Chap. 6 to such a variational reformulation, and recover, from a good approximation of it, a reasonable approximation of the optimal strategy of the original optimal control problem.

8.2 Reduction to Variational Problems

We have already envisioned how to transform an optimal control problem into a variational problem. When there are point wise constraints for admissible controls, these are reflected on certain point wise constraints for specific expressions involving states and derivatives of states. Instead of discussing a general procedure, we will be looking at some selected examples to explain this strategy. In particular, in this section we will be working with the following example.

Example 8.1 This particular case is taken from [10]. The population of a given fish species grows at a constant rate when there is no fish activity according to the law $x'(t) = x(t)$. Fish rules, however, permit a maximum fishing rate which is proportional to the size of the population expressed in the equation $x'(t) = x(t) - u(t)x(t)$ for $0 \leq u(t) \leq h$, h is the maximum possible fishing rate, and the term $u(t)x(t)$ stands precisely for the fishing rate at time t. If the horizon T is given, and the initial population is $x(0) = x_0$, determine the optimal strategy $u(t)$ to maximize the total amount of fish that can be caught.

Fig. 8.1 The fishing
problem: evolution of the
amount of fish

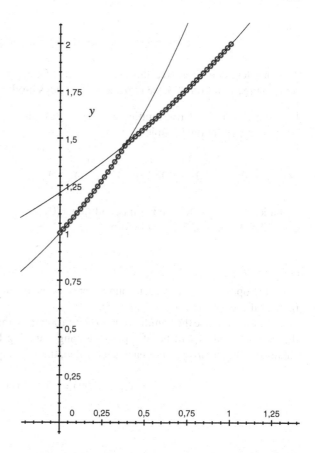

After the experience we already have with optimal control, it is straightforward
to write the problem in compact form as

$$\text{Minimize in } u(t) \in [0, h]: \quad -\int_0^T u(t)x(t)\, dt$$

subject to

$$x'(t) = (1 - u(t))x(t) \text{ in } (0, T), \quad x(0) = x_0.$$

In order to transform this problem into a variational problem, we just have to solve for
$u(t)$ into the state law, and substitute it into the integral functional. We find, bearing
in mind that $x > 0$ at all times, that the variational problem is

$$\text{Minimize in } x(t): \quad \int_0^T [x'(t) - x(t)]\, dt$$

subject to
$$x(0) = x_0, \quad (1 - h)x(t) \le x'(t) \le x(t).$$

This problem can be treated through the ideas of Chap. 6. Namely, the mechanism relies on the iteration of these two successive steps until convergence:

1. For given non-negative multipliers $y_1(t)$, $y_2(t)$, find a good approximation of the standard variational problem

$$\text{Minimize in } (x_1(t), x_2(t)) : \quad \int_0^T [x' - x + e^{y_1((1-h)x-x')} + e^{y_2(x'-x)}]\, dt$$

   under the unique left-end point condition $x(0) = x_0$.
2. Update the multipliers in the form

$$y_1 \mapsto e^{y_1((1-h)x-x')} y_1, \quad y_2 \mapsto e^{y_2(x'-x)} y_2.$$

Once the optimal $x(t)$ has been found (approximated), the optimal control is given by the difference $u(t) = 1 - x'(t)/x(t)$.

Note that because the hamiltonian for this problem is linear in the control variable, the optimal control will be of type bang-bang, jumping between the two extreme values $u = 0$, and $u = h$. For each such value, the state equations become

$$x'(t) = x(t), \quad x'(t) = (1 - h)x(t).$$

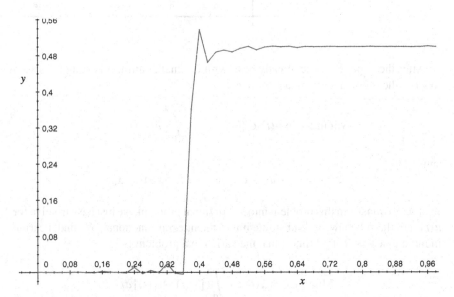

Fig. 8.2 The fishing problem: optimal control

The optimal state will therefore jump between these two families of solutions. Since we are just interested in the numerical approximation, we do not include here a full analytical discussion of this problem.

You can see, in Fig. 8.1, the evolution of the amount of fish under the action of the optimal control. We have used the specific values $T = 1$, $h = 0.5$. Figure 8.2 shows the optimal control itself. The bang-bang nature of the control is reasonably captured, and the two exponentials for the optimal evolution of the amount of fish are also well traced.

8.3 A Much More Demanding Situation

We next focus on the control of the harmonic oscillator.

Example 8.2 We pretend to lead the system

$$x''(t) + x(t) = u(t) \text{ in } (0, T), \quad |u(t)| \leq 1$$

Fig. 8.3 The harmonic oscillator for initial value $(0.5, 0.5)$: $T = 1.752$

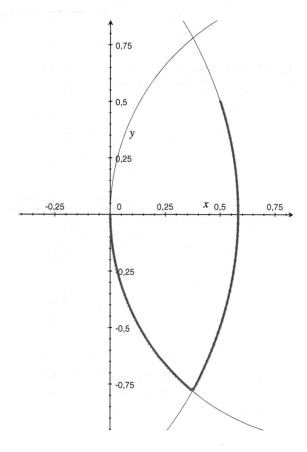

Fig. 8.4 The optimal control for the harmonic oscillator: case (0.5, 0.5)

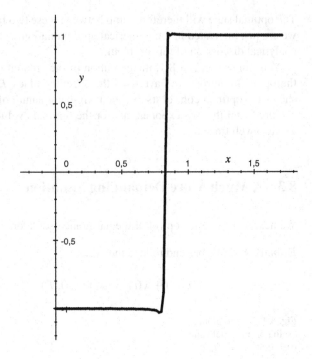

Fig. 8.5 The harmonic oscillator with extra time: case (0.5, 0.5): $T = 2$

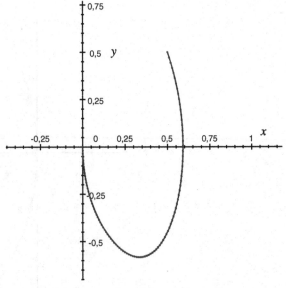

from a given initial condition (x_0, x_0') to rest $(0, 0)$ as soon as possible for a minimal time T. If we rewrite the second-order state law as a first-order system, in the standard way, we find

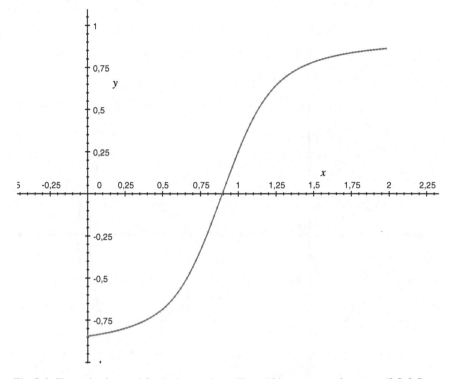

Fig. 8.6 The optimal control for the harmonic oscillator with some extra time: case $(0.5, 0.5)$

$$x_1' = x_2, \quad x_2' = -x_1 + u.$$

Suppose we fix the time T to exactly the optimal time necessary to take the system from (x_0, x_0') to $(0, 0)$. We are interested in finding the optimal solution of

$$\text{Minimize in } u(t) \in [-1, 1] : \quad \int_0^T dt$$

under

$$x_1' = x_2, \quad x_2' = -x_1 + u, \quad x_1(0) = x_0, x_2(0) = x_0', x_1(T) = 0, x_2(T) = 0.$$

If we eliminate the control variable from the formulation of the problem, we face the variational problem

$$\text{Minimize in } (x_1, x_2) : \quad \int_0^T dt$$

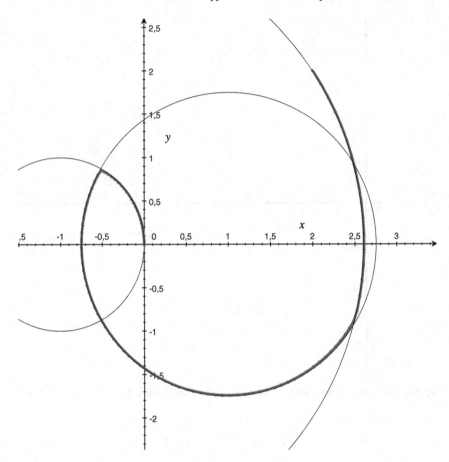

Fig. 8.7 The harmonic oscillator for initial value (2., 2.): $T = 4.9$

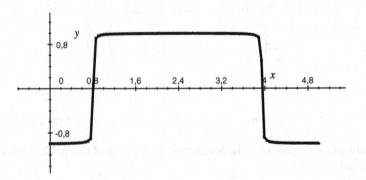

Fig. 8.8 The optimal control for the harmonic oscillator: case (2., 2.)

subject to

$$(x_1' - x_2)^2 \le 0, \quad (x_2' + x_1)^2 \le 1, \qquad x_1(0) = x_0, x_2(0) = x_0', x_1(T) = 0, x_2(T) = 0.$$

Our iterative mechanism to approximate the optimal solution introduces four non-negative multipliers y_1, y_2, y_3, y_4, and proceed, iteratively until convergence, in two steps:

1. For given $y_1(t)$, $y_2(t)$, $y_3(t)$, $y_4(t)$, find the optimal solution of the standard variational problem

$$\text{Minimize in } (x_1(t), x_2(t)) : \quad \int_0^T [1 + e^{y_1(t)(x_1'(t)-x_2(t))} + e^{y_2(t)(-x_1'(t)+x_2(t))}$$
$$+ e^{y_3(t)(x_2'(t)+x_1(t)-1)} + e^{y_4(t)(-x_2'(t)-x_1(t)+1)}] \, dt$$

under end-point conditions $x_1(0) = x_0, x_2(0) = x_0', x_1(T) = 0, x_2(T) = 0$.

2. Update rule for y_1, y_2, y_3, y_4, through formulae

$$y_1(t) \mapsto e^{y_1(t)(x_1'(t)-x_2(t))} y_1(t), \quad y_2(t) \mapsto e^{y_2(t)(-x_1'(t)+x_2(t))} y_2(t),$$
$$y_3(t) \mapsto e^{y_4(t)(x_2'(t)+x_1(t)-1)} y_3(t), \quad y_4(t) \mapsto e^{y_4(t)(-x_2'(t)-x_1(t)+1)} y_4(t),$$

See the resulting simulations by using this approach in the following figures:

- Figure 8.3 shows the optimal path from the starting point $(0.5, 0.5)$ to the origin. The two circles, as integral curves of the limit systems, have also been drawn for comparison purposes.
- Figure 8.4 shows the optimal strategy. Notice the bang-bang nature, and the abrupt jump from -1 to 1.
- Figure 8.5 shows the same situation, but we have allowed for some extra time to do the task of taking the system from $(0.5, 0.5)$ to the origin. The optimal state trajectory, as well as the optimal control (Fig. 8.6), become much smoother.
- Figure 8.7 corresponds to a much more demanding case. The starting point is $(2, 2)$. Notice how several circles coming from the two limit systems are used.
- Figure 8.8 shows the bang-bang nature of the optimal control.

Some other examples are treated in the exercises of the chapter.

8.4 Dynamic Programming

Another appealing possibility to discretize an optimal control problem is to approximate the derivative x' in the state system by a suitable difference quotient. This is an interesting viewpoint as it directly leads to a discrete optimization problem with some special features.

To be specific, suppose we are facing the optimal control problem

$$\text{Minimize in } \mathbf{u}(t): \quad \int_0^T \frac{1}{2} \left(|\mathbf{x}(t)|^2 + |\mathbf{u}(t)|^2 \right) dt$$

subjected to

$$\mathbf{x}'(t) = \mathbf{A}\mathbf{x}(t) + \mathbf{B}\mathbf{u}(t) \text{ in } (0, T), \quad \mathbf{x}(0) = \mathbf{x}_0.$$

If we divide the time interval $[0, T]$ in m subintervals of equal length $h = T/m$, and put $t_k = kT/m$ for $k = 0, 1, \ldots m$, we can use the approximation

$$\frac{1}{h}(\mathbf{x}(t_{k+1}) - \mathbf{x}(t_k)) \sim \mathbf{x}'(t_k) = \mathbf{A}\mathbf{x}(t_k) + \mathbf{B}\mathbf{u}(t_k), \quad k = 0, 1, \ldots, m - 1.$$

If we set $\mathbf{x}(t_k) \equiv \mathbf{x}_k$, $\mathbf{u}(t_k) \equiv \mathbf{u}_k$, then we have the rule

$$\mathbf{x}_{k+1} = (\mathbf{1} + h\mathbf{A})\mathbf{x}_k + h\mathbf{B}\mathbf{u}_k, \quad k = 0, 1, \ldots, m - 1, \quad \mathbf{x}_0, \text{ given.}$$

$\mathbf{1}$ is the identity matrix. Similarly, the integral cost can be approximated through a simple quadrature rule like

$$\int_0^T \frac{1}{2} \left(|\mathbf{x}(t)|^2 + |\mathbf{u}(t)|^2 \right) dt \sim \frac{h}{4} \left(|\mathbf{x}_0|^2 + |\mathbf{x}_m|^2 \right) + \frac{h}{2} \sum_{k=0}^{m-1} \left(|\mathbf{x}_k|^2 + |\mathbf{u}_k|^2 \right).$$

We have played a little bit with $\mathbf{u}_0$, in order to preserve the symmetry of the formula, since being a given, fixed vector, its size does not interfere with the optimization process. Altogether, and neglecting constant terms and fixed positive factors, the discretization of the original continuous optimal control problem would read

$$\text{Minimize in } (\mathbf{u}_k)_{k=0,1,\ldots,m-1}: \quad \frac{1}{2}|\mathbf{x}_m|^2 + \sum_{k=0}^{m-1} \left(|\mathbf{x}_k|^2 + |\mathbf{u}_k|^2 \right)$$

subjected to

$$\mathbf{x}_{k+1} = (\mathbf{1} + h\mathbf{A})\mathbf{x}_k + h\mathbf{B}\mathbf{u}_k, \quad k = 0, 1, \ldots, m - 1, \quad \mathbf{x}_0, \text{ given.}$$

This is, after all, a mathematical program stated certainly in a very special way because the cost functional is defined recursively, through intermediate variables $\mathbf{x}_k$, in terms of the independent variables $\mathbf{u}_k$. The solution of such a program would yield a nice approximation to the original optimal control problem. This is in fact one quite reasonable strategy. However, this discretization falls under the category of a much wider collection of problems which, because of the way they are stated, deserve a special treatment.

The general problem would be

$$\text{Minimize in } (\mathbf{u}_k)_{k=1,\dots,m} : \quad \sum_{k=1}^{m} F(\mathbf{x}_k, \mathbf{u}_k)$$

under

$$\mathbf{x}_k = \mathbf{f}(\mathbf{x}_{k-1}, \mathbf{u}_k) \text{ for } k = 1, 2, \dots, m, \quad \mathbf{x}_0, \text{ given.}$$

Map $\mathbf{f}(\mathbf{x}, \mathbf{u}) : \mathbb{R}^N \times \mathbb{R}^n \to \mathbb{R}^N$ and function $F(\mathbf{x}, \mathbf{u}) : \mathbb{R}^N \times \mathbb{R}^n \to \mathbb{R}$ are given. It is easy to identify these two main ingredients for our previous discretization of an optimal control problem. To realize the recursive nature of the cost functional, let us start by making computations explicit in the following way. First $\mathbf{x}_1 = \mathbf{f}(\mathbf{x}_0, \mathbf{u}_1)$, and so the first term in the cost function is $F(\mathbf{f}(\mathbf{x}_0, \mathbf{u}_1), \mathbf{u}_1)$. Likewise

$$\mathbf{x}_2 = \mathbf{f}(\mathbf{x}_1, \mathbf{u}_2) = \mathbf{f}(\mathbf{f}(\mathbf{x}_0, \mathbf{u}_1), \mathbf{u}_2),$$
$$F(\mathbf{x}_1, \mathbf{u}_1) + F(\mathbf{x}_2, \mathbf{u}_2) = F(\mathbf{f}(\mathbf{x}_0, \mathbf{u}_1), \mathbf{u}_1) + F(\mathbf{f}(\mathbf{f}(\mathbf{x}_0, \mathbf{u}_1), \mathbf{u}_2), \mathbf{u}_2),$$

and so on. It must be clear that this is not a practical way to proceed. Some other strategy needs to be organized.

Let us look at the first step in a different way. Suppose $\mathbf{x}_1$ is given and fixed (in addition to $\mathbf{x}_0$), and look at the optimization problem

$$\text{Minimize in } \mathbf{u} \in \mathbb{R}^n : \quad F(\mathbf{x}_1, \mathbf{u})$$

subjected to

$$\mathbf{x}_1 = \mathbf{f}(\mathbf{x}_0, \mathbf{u}).$$

This is a much easier mathematical program to be solved in the only variable $\mathbf{u}$ under N equality constraints. Let us just denote, for record purposes, by $\mathscr{F}_1(\mathbf{x}_1)$ the value of this minimum as a function of the state $\mathbf{x}_1$, regarded now as an independent moving variable. There is an optimal $\mathbf{u} \equiv \mathbf{u}_1(\mathbf{x}_1)$ where that minimum is attained. How would we set up the second step? For given $\mathbf{x}_2$, we would be interested in the problem

$$\text{Minimize in } (\mathbf{x}, \mathbf{u}) \in \mathbb{R}^N \times \mathbb{R}^n : \quad \mathscr{F}_1(\mathbf{x}) + F(\mathbf{x}_2, \mathbf{u})$$

under

$$\mathbf{x}_2 = \mathbf{f}(\mathbf{x}, \mathbf{u}).$$

Let $\mathscr{F}_2(\mathbf{x}_2)$ denote this minimum value, and $\mathbf{u} \equiv \mathbf{u}_2(\mathbf{x}_2)$, $\mathbf{x} \equiv \mathbf{x}_1(\mathbf{x}_2)$, an optimal pair realizing that minimum. Before writing down the general law, let us focus on one more step: for each $\mathbf{x}_3$, solve the problem

$$\text{Minimize in } (\mathbf{x}, \mathbf{u}) \in \mathbb{R}^N \times \mathbb{R}^n : \quad \mathscr{F}_2(\mathbf{x}) + F(\mathbf{x}_3, \mathbf{u})$$

subject to

$$\mathbf{x}_3 = \mathbf{f}(\mathbf{x}, \mathbf{u}),$$

and put $\mathscr{F}_3(\mathbf{x}_3)$ for the minimum value, as a function of $\mathbf{x}_3$, and identify at least one optimal pair $\mathbf{u} = \mathbf{u}_3(\mathbf{x}_3)$, $\mathbf{x} = \mathbf{x}_2(\mathbf{x}_3)$. If our original problem would have three steps

$$\text{Minimize in } (\mathbf{u}_1, \mathbf{u}_2, \mathbf{u}_3) \in \mathbb{R}^n \times \mathbb{R}^n \times \mathbb{R}^n : \quad F(\mathbf{x}_1, \mathbf{u}_1) + F(\mathbf{x}_2, \mathbf{u}_2) + F(\mathbf{x}_3, \mathbf{u}_3)$$

subjected to

$$\mathbf{x}_1 = \mathbf{f}(\mathbf{x}_0, \mathbf{u}_1), \quad \mathbf{x}_2 = \mathbf{f}(\mathbf{x}_1, \mathbf{u}_2), \quad \mathbf{x}_3 = \mathbf{f}(\mathbf{x}_2, \mathbf{u}_3),$$

then it is clear that an optimal solution can be constructed backwards by finding an optimal solution $\mathbf{x} = \mathbf{x}_3$ for the final problem

$$\text{Minimize in } \mathbf{x} \in \mathbb{R}^N : \quad \mathscr{F}_3(\mathbf{x}).$$

Indeed, an optimal solution would be

$$\mathbf{x}_3, \quad \mathbf{u}_3 \equiv \mathbf{u}_3(\mathbf{x}_3), \mathbf{x}_2 \equiv \mathbf{x}_2(\mathbf{x}_3), \quad \mathbf{u}_2 \equiv \mathbf{u}_2(\mathbf{x}_2), \mathbf{x}_1 \equiv \mathbf{x}_1(\mathbf{x}_2), \quad \mathbf{u}_1 \equiv \mathbf{u}_1(\mathbf{x}_1).$$

The whole point of dynamic programming is that the problem can be solved by attacking each individual step at a time, and proceed forward until the last step.

Principle 8.1 *Suppose we have solved the j-th step by calculating $\mathscr{F}_j(\mathbf{x})$ and finding optimal maps $\mathbf{u}_j(\mathbf{x})$ and $\mathbf{x}_{j+1}(\mathbf{x})$ realizing that optimal value. Then the $j + 1$-th step is the result of the following optimization problem, for each given $\mathbf{x}_{j+1}$,*

$$\text{Minimize in } (\mathbf{x}, \mathbf{u}) \in \mathbb{R}^N \times \mathbb{R}^n : \quad \mathscr{F}_j(\mathbf{x}) + F(\mathbf{x}_{j+1}, \mathbf{u})$$

under

$$\mathbf{x}_{j+1} = \mathbf{f}(\mathbf{x}, \mathbf{u}).$$

To make the previous discussion as neat as possible, we have avoided explicit constraints on feasible control $\mathbf{u}$. Most of the time such constraints $\mathbf{u}(t) \in \mathbf{K}$, where $\mathbf{K}$ is a convex, compact subset of $\mathbb{R}^N$, would be an important part of the optimization process. Under such explicit conditions the whole process becomes much more complex because those constraints propagate to successive states x_j. To see this point more clearly, suppose we go back to the first step of the dynamic programming strategy

$$\text{Minimize in } \mathbf{u} \in \mathbf{K}: \quad F(\mathbf{x}_1, \mathbf{u})$$

subjected to

$$\mathbf{x}_1 = \mathbf{f}(\mathbf{x}_0, \mathbf{u}).$$

The issue is that $\mathbf{x}_1$ cannot be an arbitrary vector as it must belong to the image set $\mathbf{f}(\mathbf{x}_0, \mathbf{K})$. Unfortunately, these constraint on successive states $\mathbf{x}_j$ become more and more involved, and quite often they cannot be easily made explicit.

It is time to look at some simple explicit examples to understand what we have explained so far.

Example 8.3 We seek to

$$\text{Minimize in } (u_1, u_2, u_3): \quad \sum_{k=1}^{3} (x_k^2 + 2u_k^2)$$

subject to

$$x_k = x_{k-1} - u_k \text{ for } k = 1, 2, 3, \quad x_0 = 5.$$

We proceed in four steps.

1. Determine $\mathscr{F}_1(x_1)$ and optimal map $u_1(x_1)$:

$$\text{Minimize in } u: \quad x_1^2 + 2u^2 \quad \text{for} \quad x_1 = 5 - u.$$

   This is easy:

$$\mathscr{F}_1(x) = x^2 + 2(5 - x)^2, \quad u_1(x) = 5 - x.$$

2. Determine $\mathscr{F}_2(x_2)$, and optimal maps $u_2(x_2)$, $x_1(x_2)$:

$$\text{Minimize in } (x, u): \quad \mathscr{F}_1(x) + x_2^2 + 2u^2 \quad \text{for} \quad x_2 = x - u.$$

   It is elementary to see that

$$\mathscr{F}_2(x) = \frac{11}{5}x^2 - 8x + 30, \quad x_1(x) = 2 + \frac{2}{5}x, u_2(x) = 2 - \frac{3}{5}x.$$

3. Find $\mathscr{F}_3(x_3)$ and optimal maps $x_2(x_3)$, $u_3(x_3)$ by solving

$$\text{Minimize in } (x, u): \quad \mathscr{F}_2(x) + x_3^2 + 2u^2 \quad \text{for} \quad x_3 = x - u.$$

Again, it is easy to find the optimal solution

$$\mathscr{F}_3(x_3) = \frac{1}{21}(43x_3^2 - 80x_3 + 550), \quad x_2(x) = \frac{10}{21}(2+x), u_3(x) = \frac{1}{21}(20 - 11x).$$

4. The last step is to find the optimal solution of

$$\text{Minimize in } x_3 : \quad \mathscr{F}_3(x_3).$$

The optimal solution is $x_3 = 40/43$. From here we proceed backwards to have

$$u_3 = u_3(40/43) = 20/43, \quad x_2 = x_2(40/43) = 60/43,$$
$$u_2 = u_2(60/43) = 50/43, \quad x_1 = x_1(60/43) = 110/43, u_1 = u_1(110/43) = 105/43.$$

This is an interesting example where one can see and compare the perspective we have tried to implement in this text concerning the numerical approximation. Our philosophy is to eliminate the u-variables by substituting

$$u_1 = 5 - x_1, \quad u_2 = x_1 - x_2, \quad u_3 = x_2 - x_3$$

into the cost, and deal directly with a problem in the x-variables, namely,

$$\text{Minimize in } (x_1, x_2, x_3) : \quad x_1^2 + 2(5 - x_1)^2 + x_2^2 + 2(x_1 - x_2)^2 + x_3^2 + 2(x_2 - x_3)^2,$$

under no constraint. It is very easy to write down the linear system furnishing the optimal solution which is

$$(x_1, x_2, x_3) = \frac{1}{43}(110, 60, 40),$$

and from here

$$(u_1, u_2, u_3) = \frac{1}{43}(105, 50, 20).$$

Example 8.4 If we put explicit constraints into the problem like

$$\text{Minimize in } (u_1, u_2, u_3) \in [0, 2] : \quad \sum_{k=1}^{3}(x_k^2 + 2u_k^2)$$

subject to

$$x_k = x_{k-1} - u_k \text{ for } k = 1, 2, 3, \quad x_0 = 5,$$

then it is even harder to implement by naked hands the dynamic programming approach. It is, however, worth trying. Our viewpoint leads to deal with the problem

$$\text{Minimize in } (x_1, x_2, x_3): \quad x_1^2 + 2(5 - x_1)^2 + x_2^2 + 2(x_1 - x_2)^2 + x_3^2 + 2(x_2 - x_3)^2,$$

under

$$0 \leq 5 - x_1 \leq 2, \quad 0 \leq x_1 - x_2 \leq 2, \quad 0 \leq x_2 - x_3 \leq 2.$$

More explicitly, we seek to

$$\text{Minimize in } (x_1, x_2, x_3): \quad 5x_1^2 - 20x_1 - 4x_1x_2 + 5x_2^2 - 4x_2x_3 + 3x_3^2 + 50$$

subject to

$$3 \leq x_1 \leq 5, \quad x_1 - 2 \leq x_2 \leq x_1, \quad x_2 - 2 \leq x_3 \leq x_2.$$

There is a further relevant situation where variables u can only take on a finite, or discrete set of values, even depending on the step of the process. In the context of the previous example that would correspond to the situation

$$\text{Minimize in } (u_1, u_2, u_3) \in \{0, 1, 2\}^3: \quad \sum_{k=1}^{3}(x_k^2 + 2u_k^2)$$

subject to

$$x_k = x_{k-1} - u_k \text{ for } k = 1, 2, 3, \quad x_0 = 5,$$

or even to

$$\text{Minimize in } (u_1, u_2, u_3) \in \{0, 1, 2\} \times \{1, 2\} \times \{2\}: \quad \sum_{k=1}^{3}(x_k^2 + 2u_k^2)$$

subject to

$$x_k = x_{k-1} - u_k \text{ for } k = 1, 2, 3, \quad x_0 = 5,$$

Instead of discussing this particular example, we focus on a more applied and appealing typical situation.

Example 8.5 The following table gives the cumulative demand of power supply plants for the next six years, and the estimated price (in the appropriate units) for each plant and each year.[1] There is also a unique payment of 1500 for each year in which some plants are decided to be built regardless of how many.

---

[1] From http://web.mit.edu/15.053/www/AMP-Chapter-11.pdf with some changes on the data set.

Year	Cumulative demand	Cost per plant
1	1	5400
2	2	5600
3	4	5000
4	6	5100
5	7	5000
6	8	5200

The optimal (less expensive) construction plan is sought.

The problem proceeds in six steps in which we have to decide how many plants are to be built, ensuring that the cumulative demand is met for each such step. Dynamic programming proceeds one step at a time in the following way.

1. For stage 1, there is no much to be discussed. We can construct from 1 to 8 plants, and the cost $c_1(n)$ would be $n \times 5400 + 1500$.
2. For step 2, we can have seven possible states, $n = 2, 3, \ldots, 8$, as we need to ensure to have at least 2 plants working. The cost of each such possibility $n \geq 2$ is

$$c_2(n) = \min\{ \min_{j \in \{1,\ldots,n-1\}} \{c_1(n-j) + j \times 5600 + 1500\}, c_1(n)\}.$$

It is easy to check that

$$c_2(n) = c_1(n) = 5400 \times n + 1500,$$

all plants are built in the first step, and none in the second.
3. For the third stage, we start at $n = 4$ because we have to ensure at least this number of plants working. Then, just as before,

$$c_3(n) = \min\{ \min_{j \in \{1,\ldots,n-2\}} \{c_2(n-j) + j \times 5000 + 1500\}, c_2(n)\}, \quad n \geq 4.$$

Since we have an explicit formula for $c_2(k)$ form the preceding step, we can see that

$$c_3(n) = \begin{cases} 5400 \times n + 1500, & n = 4, 5, \\ 5000 \times n + 3800, & n = 6, 7, 8. \end{cases}$$

4. Once again, for $n = 6, 7, 8$, we find that

$$c_4(n) = \min\{ \min_{j \in \{1,\ldots,n-4\}} \{c_3(n-j) + j \times 5000 + 1500\}, c_2(n)\}, \quad n \geq 6.$$

Since the formula for the optimal value $c_3(k)$ depends on k, we evaluate $c_4(n)$ individually for each $n = 6, 7, 8$. In fact,

$$c_4(6) = \min\{c_3(5) + 6600, c_3(4) + 11700, 33800\} = 33800,$$
$$c_4(7) = \min\{c_3(6) + 6600, c_3(5) + 11700, 38800\} = 38800,$$
$$c_4(8) = \min\{c_3(7) + 6600, c_3(6) + 11700, c_3(5) + 16800, c_3(8)\} = c_3(8) = 43800.$$

5. For the fifth stage, we would have

$$c_5(7) = \min\{c_4(7), c_4(6) + 6500\} = 38800,$$
$$c_5(8) = \min\{c_4(8), c_4(7) + 6500, c_4(6) + 11500\} = c_4(8) = 43800.$$

6. Finally

$$c_6(8) = \min\{c_5(8), c_5(7) + 6700\} = c_5(8) = 43800.$$

We therefore conclude that the minimal cost is 43800 unitary units, according to the following construction plan

$$2, 0, 6, 0, 0, 0.$$

These values are easily obtained reviewing backwards the various stages and checking where the successive minima were attained.

Though this example is pretty elementary, it is a nice situation where the basic principle of discrete dynamic programming can be seen in practice. If we have that a certain situation evolves along several stages $k = 1, 2, \ldots, N$ through successive states $\mathbf{x} \in \mathbf{A}_k$, and passage from one step $\mathbf{x} \in \mathbf{A}_{k-1}$ to the next $\mathbf{y} \in \mathbf{A}_k$ involves a precise cost $c(k, \mathbf{x}, \mathbf{y})$ depending on the step, the starting state $\mathbf{x}$ and the final state $\mathbf{y}$, we would like to know the optimal global evolution through the various stages minimizing the global price

$$\sum_{k=1}^{N} c(k, \mathbf{x}_{k-1}, \mathbf{x}_k)$$

if the system is expected to end at a certain state $\mathbf{X}$ in the final stage. For $k = 1$, $c(1, \mathbf{x}_0, \mathbf{x}_1)$ simply means $c(1, \mathbf{x}_1)$. The optimization problem would read

$$\text{Minimize in } (\mathbf{x}_k)_{k=1,2,\ldots,N} : \quad \sum_{k=1}^{N} c(k, \mathbf{x}_{k-1}, \mathbf{x}_k)$$

subject to

$$\mathbf{x}_k \in \mathbf{A}_k, k < N, \quad \mathbf{x}_N = \mathbf{X}.$$

The basic idea of dynamic programming to find the optimal solution is to proceed one step at a time successively by means of the principle of dynamic programming.

Principle 8.2 *If we let, for* $\mathbf{x} \in \mathbf{A}_j$, $c_j(\mathbf{x})$ *stand for the optimal value of the problem*

$$\text{Minimize in } (\mathbf{x}_k)_{k=1,2,\ldots,j} : \quad \sum_{k=1}^{j} c(k, \mathbf{x}_{k-1}, \mathbf{x}_k)$$

subject to

$$\mathbf{x}_k \in \mathbf{A}_k, k < j, \quad \mathbf{x}_j = \mathbf{x},$$

then, for $\mathbf{x} \in \mathbf{A}_{j+1}$,

$$c_{j+1}(\mathbf{x}) = \min_{\mathbf{y} \in \mathbf{A}_j} \{c_j(\mathbf{y}) + c(j+1, \mathbf{y}, \mathbf{x})\}.$$

This is precisely the strategy used in the previous example.

8.5 Final Remarks

This technique of transforming an optimal control problem into an equivalent variational problem is always possible. Indeed, if we look at the problem

$$\text{Minimize in } \mathbf{u}(t) \in \mathbf{K} : \quad \int_0^T F(\mathbf{x}(t), \mathbf{u}(t)) \, dt$$

under

$$\mathbf{x}(t)' = \mathbf{f}(\mathbf{x}(t), \mathbf{u}(t)) \text{ in } (0, T), \quad \mathbf{x}(0) = \mathbf{x}_0, (\mathbf{x}(T) = \mathbf{x}_T), \quad \mathbf{x}(t) \in \Omega,$$

and define the integrand

$$\psi(\mathbf{x}, \xi) = \min_{\mathbf{u}} \{F(\mathbf{x}, \mathbf{u}) : \mathbf{u} \in \mathbf{K}, \xi = \mathbf{f}(\mathbf{x}, \mathbf{u})\},$$

then it turns out that the optimal control problem is equivalent to the variational problem

$$\text{Minimize in } \mathbf{x}(t) : \quad \int_0^T \psi(\mathbf{x}(t), \mathbf{x}'(t)) \, dt$$

subject to $\mathbf{x}(0) = \mathbf{x}_0, (\mathbf{x}(T) = \mathbf{x}_T)$, and $\mathbf{x}(t) \in \Omega$ [20]. In this text, we have restricted our attention to simple situations where the integrand ψ can be written explicitly in a simple way, essentially because the identity $\xi = \mathbf{f}(\mathbf{x}, \mathbf{u})$ allows to solve in a neat way for $\mathbf{u}$ in terms of $(\mathbf{x}, \xi)$. When this is not the case, the approximation is much more involved but still possible. This is however beyond the scope of this text.

There are quite sophisticated simulation methods for optimal control. They produce very accurate approximations but require a good deal of expertise, as already pointed out. See the bibliography section for further references.

8.6 Exercises

8.6.1 Exercises to Support the Main Concepts

1. Compare the two optimal control problems

$$\text{Minimize in } u(t): \quad \int_0^1 \frac{1}{2} u(t)^2 \, dt$$

under

$$x'(t) = x(t) + u(t) \text{ in } (0, 1), \quad x(0) = 1, x(1) = 0,$$

and

$$\text{Minimize in } u(t): \quad \int_0^1 \frac{1}{2} u(t)^2 \, dt$$

under

$$x'(t) = x(t) + u(t) \text{ in } (0, 1), |u(t)| \le 1, \quad x(0) = 1, x(1) = 0.$$

by identifying the corresponding integrand for the variational reformulation for both.
2. Write, as explicitly as possible, the integrand for the equivalent variational problem corresponding to

$$\text{Minimize in } u(t): \quad \frac{1}{2} \int_0^1 u(t)^2 \, dt$$

under

$$x'(t) = u(t)x(t) + u(t)^2 \text{ in } (0, 1), \quad x(0) = 0, x(1) = 0.5.$$

8.6.2 Computer Exercises

1. Consider the optimal control problem

$$\text{Minimize in } u(t): \quad \frac{1}{2} x^2(T) + \int_0^T \left(\frac{3}{2} x(t)^2 + \frac{1}{2} u(t)^2 \right) \, dt$$

under

$$x'(t) = x(t) + u(t) \text{ in } (0, T), \quad x(0) = x_0.$$

(a) Find the optimal solution.
(b) For the particular values $T = 2$, $x_0 = 1$, find an approximation, and compare it with the exact solution.

2. For an oscillator with friction obeying the equation

$$x''(t) + \alpha x'(t) + x(t) = u(t), \quad |u(t)| \le 1,$$

find an approximation of some optimal curves for $\alpha = 0.4$, and compare it to the non-friction case $\alpha = 0$.

(a) Take $\alpha = 0$ (no friction), and find your own approximation to the optimal minimum-time curve taking the system from $(2, 2)$ to $(0, 0)$ (this has been shown in the chapter).
(b) Using that minimal time, compute the optimal curves for $\alpha = 0.4$ to see the effect of friction.
(c) Find the minimum optimal time for the friction case $\alpha = 0.4$.

3. For the optimal control problem

$$\frac{1}{2} \int_0^{10} [\beta x_1(t)^2 + r u(t)^2] \, dt$$

with state law

$$x_1'(t) = x_2(t), \quad x_2'(t) = -x_2(t) + u(t)$$

and end-point conditions

$$x_1(0) = x_2(0) = 1, \quad x_1(10) = x_2(10) = 0,$$

approximate the optimal solution for the cases:

(a) $\beta = 0, r = 0.1$;
(b) $\beta = 1, r = 0.1$;
(c) $\beta = 1, r = 1$.

4. For a double integrator $x''(t) = u(t)$, take the cost to be

$$\int_0^T \frac{1}{2} \left(x'(t)^2 + u(t)^2 \right) \, dt.$$

Approximate the optimal use of the accelerator if $x(0) = x'(0) = 0$, $x(T) = L$ but $x'(T)$ free. Particular values are $T = 1$, $L = 1$.

5. Duffing's equation

$$x''(t) + 0.4x'(t) + x(t) + 0.3x(t)^3 = u(t)$$

is a non-linear (cubic) perturbation of the equation for a linear harmonic oscillator with friction. By means of the control u, $|u| \leq 1$, we would like to lead the system from $(1.5, 1.5)$ to rest $(0, 0)$. Take $T = 7$. Compare the effect of the cubic term for $T = 2.55$, and without that term (linear case) with $T = 2.8$.

6. To approximate the minimal time in which a certain task can be accomplished is a challenging task. One possibility is to use a bisection method on the time horizon until a sufficiently accurate estimate is found. For the classical harmonic oscillator

$$x''(t) + x(t) = u(t), \quad |u(t)| \leq 1,$$

departing from $x(0) = (2, 2)$, we would like to approximate the minimal time T in which $x(T) = (0, 0)$. We already know that $T \sim 5.05$, but let us pretend not to have this information.

(a) Star with $T = 8$, and find the optimal signal $u_{10}(t)$, and optimal path. The algorithm should converge without any difficulty: it means that the task can be accomplished in $T = 8$ units of time.

(b) Divide $T = 8$ by half, and try out with $T = 4$. What do you observe? This time the algorithm cannot converge as the job we pretend to perform is impossible.

(c) Take now the middle point between 8 and 4, $T = 6$. Keep doing this until you see a bang-bang behavior, so that you find a reasonable approximation of the minimal time.

8.6.3 Case Studies

1. In the context of the gout problem in the previous chapter, take as cost functional

$$\frac{1}{2} \int_0^T u(t)^2 \, dt,$$

state law $x'(t) = -x(t) + 1 - u(t), 0 \leq u(t) \leq 2, x(0) = 1, x(T) = 0$. Approximate the optimal solution for several values of T. Comment on the behavior of the optimal dosage plan as T becomes smaller.

2. An autopilot for a ship is in charge of maintaining the ship's heading straight.[2] A simple model is given by the equations

$$\theta''(t) = -\theta'(t) - \alpha(t), \quad \alpha'(t) = -\alpha(t) + u(t),$$

---

[2]From [5] with changed data set.

where θ is the heading error (the angle between the ship's true and the desired headings), α is the rudder angle, and u is the commanded rudder angle (the control). M, d and c are parameters. If initially we find that $\theta = 0.1$, $\theta' = 0.05$, and $\alpha = 0$, find the optimal use of the commanded rudder angle u, provided that $|u| \leq 1$, to bring the ship to $\theta = \theta' = 0$ in $T = 8$ and $T = 5$ units of time, while leaving $\alpha = 0$, as well.

3. A farmer produces four dairy products: cheese, butter, cottage cheese, and yogurt. The benefit coming from using $1, 2, 3, 4$ or 5 units of milk is shown in the following table:

Units of milk	1	2	3	4	5
Cheese	9	17	23	29	32
Butter	5	11	15	18	21
Cottage cheese	8	13	15	16	17
Yogurt	7	14	20	23	25

By using dynamic programming, seek the amounts of the four products to be produced with 5 units of milk so as to maximize benefits.

4. The city hall in a town has received 5 million euros from the goverment to promote welfare in areas like Youth, Culture and Environment. The analysis team has estimated the level of satisfaction of citizens en terms of the distribution of budget (multiples of a million) according to the following table:

Area/Amount	1	2	3	4	5
Youth	5	5	6	7	11
Culture	3	6	6	8	10
Environment	5	6	8	9	14

Seek, by using the dynamic programming ideas, the optimal way to divide the total budget to maximize the level of the satisfaction of population.

5. The state of a flexible arm for a robot is governed by the system

$$
\mathbf{x}'(t) = \begin{pmatrix} 0 & 1 & 0 & 0 \\ -c/Jn^2 & -v/J & c/Jn^2 & 0 \\ 0 & 0 & 0 & 1 \\ c/ml^2 & 0 & -c/ml^2 & 0 \end{pmatrix} \mathbf{x}(t) + \begin{pmatrix} 0 \\ k/Jn \\ 0 \\ 0 \end{pmatrix} u(t),
$$

where realistic values of the parameters are:

- $m = 0.04371$ Kg.;
- $l = 0.98$ m.;
- $c = 0.377$ Nm.;
- -1.8 V. $\leq u \leq +1.7$ V.;
- $k = 0.21$ Nm/V.;

Fig. 8.9 A flexible arm for a robot

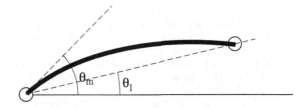

- $J = 1.47 \times 10^{-4}$ Kg s^2;
- $\nu = 1.05 \times 10^{-3}$ Kg m^2 g;
- $n = 50$.

The four components of the state vector are

$$\mathbf{x} = (\theta_m, \theta'_m, \theta_l, \theta'_l)$$

where the two angles θ_m, θ_l are represented in Fig. 8.9. The task consists in moving the arm from the position $\theta_m = \theta_l = 0$ at rest to $\theta_m = \theta_l = \pi/2$, again at rest, for several time horizons: $T = 2$, $T = 1.7$, $T = 1.4$, $T = 1.2$, $T = 1.125$. What happens if you push up to $T = 1.1$?

References

1. S. Anita, V. Arnautu, V. Capasso, *An Introduction to Optimal Control Problems in Life Sciences and Economics. From Mathematical Models to Numerical Simulation with MATLAB*, Modeling and Simulation in Science, Engineering and Technology (Birkhäuser/Springer, New York, 2011)
2. L.D. Berkovitz, N.G. Medhin, *Nonlinear Optimal Control Theory. With a Foreword by Wendell Fleming*, Chapman & Hall/CRC Applied Mathematics and Nonlinear Science Series (CRC Press, Boca Raton, FL, 2013)
3. J.T. Betts, *Practical Methods for Optimal Control and Estimation Using Nonlinear Programming*, 2nd edn. Advances in Design and Control, Society for Industrial and Applied Mathematics (SIAM), vol. 19 (Philadelphia, PA, 2010)
4. J. Blot, N. Hayek, *Infinite-Horizon Optimal Control in the Discrete-Time Framework*, Springer-Briefs in Optimization (Springer, New York, 2014)
5. J.B. Burl, *Linear Optimal Control*, Addison Wesley Longman Inc. (Menlo Park, California, 1999)
6. J.A. Burns, *Introduction to the Calculus of Variations and Control with Modern Applications*, Chapman & Hall/CRC Applied Mathematics and Nonlinear Science Series (CRC Press, Boca Raton, FL, 2014)
7. P. Chen, S.M.N. Islam, *Optimal Control Models in Finance. A New Computational Approach*, Applied Optimization, vol. 95 (Springer-Verlag, New York, 2005)
8. F. Clarke, *Functional Analysis, Calculus of Variations and Optimal Control*, Graduate Texts in Mathematics, vol. 264 (Springer, London, 2013)
9. D. Grass, J.P. Caulkins, G. Feichtinger, G. Tragler, D.A. Behrens, *Optimal Control of Nonlinear Processes. With Applications in Drugs, Corruption, and Terror* (Springer-Verlag, Berlin, 2008)

10. L.M. Hocking, *Optimal Control, An Introduction to the Theory with Applications, Oxford Applied Mathematics and Computing Science Series* (Clarendon Press, Oxford, 1997)
11. B. de Jager, T. van Keulen, J. Kessels, *Optimal Control of Hybrid Vehicles*, Advances in Industrial Control (Springer, London, 2013)
12. S. Lenhart, J.T. Workman, *Optimal Control Applied to Biological Models*, Chapman & Hall/CRC Mathematical and Computational Biology Series (Chapman & Hall/CRC, Boca Raton, FL, 2007)
13. M. Levi, *Classical Mechanics with Calculus of Variations and Optimal Control. An Intuitive Introduction*, Student Mathematical Library, American Mathematical Society, Providence, RI; Mathematics Advanced Study Semesters, vol. 49 (University Park, PA, 2014)
14. F.L. Lewis, D.L. Vrabie, V.L. Syrmos, *Optimal Control*, 3rd edn. (John Wiley & Sons Inc, Hoboken, NJ, 2012)
15. D. Liberzon, *Calculus of Variations and Optimal Control Theory*, A Concise Introduction (Princeton University Press, Princeton, NJ, 2012)
16. J.M. Longuski, J.J. Guzmán, J.E. Prussing, *Optimal Control with Aerospace Applications*, Space Technology Library (Microcosm Press, El Segundo, CA; and Springer, New York, 2014)
17. C.R. MacCluer, *Calculus of Variations. Mechanics, Control, and Other Applications* (Pearson Prentice Hall, Upper Saddle River, NJ, 2005)
18. M. Mesterton-Gibbons, *A Primer on the Calculus of Variations and Optimal Control Theory*, Student Mathematical Library, vol. 50 (American Mathematical Society, Providence, RI, 2009)
19. P. Pedregal, *Introduction to Optimization*, Texts in Applied Mathematics, vol. 46 (Springer, 2003)
20. P. Pedregal, On the generality of variational principles. Milan J. Math. **71**, 319–356 (2003)
21. E.R. Pinch, *Optimal Control and the Calculus of Variations*, Oxford Science Publications (The Clarendon Press, Oxford University Press, New York, 1993)
22. J.L. Speyer, D.H. Jacobson, *Primer on Optimal Control Theory*, Advances in Design and Control, Society for Industrial and Applied Mathematics (SIAM), vol. 20 (Philadelphia, PA, 2010)
23. A. Takayama, *Mathematical Economics*, 2nd edn. (Cambridge University Press, 1985)
24. T.A. Weber, *Optimal Control Theory with Applications in Economics*, With a Foreword by A.V Kryazhimskiy (MIT Press, Cambridge, MA, 2011)
25. J.L. Troutman, *Variational Calculus and Optimal Control. With the Assistance of William Hrusa, Optimization with Elementary Convexity*, 2nd edn. Undergraduate Texts in Mathematics (Springer-Verlag, New York, 1996)

Part IV
Appendix

Chapter 9
Hints and Solutions to Exercises

9.1 Chapter 2

9.1.1 Exercises to Support the Main Concepts

1. According to the fundamental property of LP, it suffices to evaluate each linear function on the four vertices, and conclude on the maximum and the infimum. It is, by no means, necessary to have a full description of the feasible set, though it is not difficult (a bit tedious though) to do so.

2. To draw the feasible set of this exercise, one needs to go through the three lines

$$-x_1 + x_2 = 2, \quad 4x_1 - x_2 = -1, \quad 2x_1 + x_2 = 0,$$

and decide on each case which semi-plane corresponds to the appropriate inequality. The feasible set is then easily represented. Moreover the iso-cost lines $x_1 + 3x_2 = c$ can be drawn too. Notice that the third constraint is redundant: it does not restrict the feasible set in a way that it is not restricted by the others.

(a) It is then easy to see that the optimal solution corresponds to the vertex $(1/3, 7/3)$ with a maximum value $22/3$.

(b) We examine the three problems successively. For the first one, it is clear, keeping in mind the picture of the feasible set, that the maximum corresponds to the intersection point of the two lines

$$-x_1 + x_2 = 2 + h, \quad 4x_1 - x_2 = -1.$$

The maximum is $(22/3) + h(13/3)$. Concerning the second one, the maximum will be attained at the intersection vertex of the two lines

$$-x_1 + x_2 = 2, \quad 4x_1 - x_2 = -1 + h.$$

© Springer International Publishing AG 2017
P. Pedregal, *Optimization and Approximation*, UNITEXT 108,
DOI 10.1007/978-3-319-64843-9_9

The maximum is $(22/3)+h(4/3)$. Finally, for the third one it is clear, taking into account the picture of the admissible set, that the maximum is the same that has already been computed in the previous item $22/3$ for all small h.

(c) It is immediate to have that those derivatives are $(13/3, 4/3, 0)$.

(d) The dual problem is

$$\text{Minimize in } (y_1, y_2, y_3): \quad 2y_1 - y_2$$

under

$$-y_1 + 4y_2 - 2y_3 \geq 1, \quad y_1 - y_2 - y_3 \geq 3, \quad y_1, y_2, y_3 \geq 0.$$

At optimality, the following identities must hold

$$y_1(-x_1 + x_2 - 2) = y_2(4x_1 - x_2 + 1) = y_3(2x_1 + x_2) = 0,$$
$$x_1(-y_1 + 4y_2 - 2y_3 - 1) = x_2(y_1 - y_2 - y_3 - 3) = 0.$$

It is straightforward to conclude that

$$-y_1 + 4y_2 - 2y_3 - 1 = y_1 - y_2 - y_3 - 3 = y_3 = 0,$$

and the optimal solution for the dual is $(13/3, 4/3, 0)$, exactly the gradient of the maximum for the primal with respect to the parameter h.

3. In the new variables, the primal problem is

$$\text{Maximize in } (X_1, X_2): \quad \frac{22}{3} - \frac{13}{3}X_2 - \frac{4}{3}X_2 \text{ under } 2X_1 + X_2 \leq 3, X_1, X_2 \geq 0.$$

The dual problem (one version of it) is

$$\text{Minimize in } Y: \quad 3Y \text{ under } 2Y \geq -\frac{13}{3}, Y \geq -\frac{4}{3}, Y \geq 0,$$

whose solution is evidently $Y = 0$ with a vanishing minimum. Duality implies then that the optimal solution for the primal corresponds to $X_1 = X_2 = 0$.

4. If we put $\mathbf{X} = (\mathbf{x}_+, \mathbf{x}_-)$ so that $\mathbf{x} = \mathbf{x}_+ - \mathbf{x}_-$, and $\mathbf{X} \geq \mathbf{0}$, in terms of this new set of variables $\mathbf{X}$, the problem reads

$$\text{Maximize in } \mathbf{X}: \quad (\mathbf{c}, -\mathbf{c}) \cdot \mathbf{X} \text{ subject to } (\mathbf{A}, -\mathbf{A})\mathbf{X} \leq \mathbf{b}, \mathbf{X} \geq \mathbf{0}.$$

The dual of this problem is

$$\text{Minimize in } \mathbf{y}: \quad \mathbf{b} \cdot \mathbf{y} \text{ subject to } \mathbf{y}(\mathbf{A}, -\mathbf{A}) \leq (\mathbf{c}, -\mathbf{c}), \mathbf{y} \geq \mathbf{0}$$

which is the same as

$$\text{Minimize in } \mathbf{y}: \quad \mathbf{b} \cdot \mathbf{y} \text{ subject to } \mathbf{yA} = \mathbf{c}, \mathbf{y} \geq \mathbf{0}.$$

5. The only special aspect of this problem is that the dual problem has infinitely many solutions, and so does the primal. But otherwise, there is no particular difficulty.

6. One of the constraints of for the dual problem is $-y_1 - y_2 \geq 1$ which is impossible if $y_1, y_2 \geq 0$.

7. For h negative, it is easy to check that the feasible set is the triangle with vertices $(0,0)$, $(0,1)$, and $(-h/(1+h^2), 1/(1+h^2))$. The three values of the cost in these three vertices are 0, -7, and $-(2h+7)/(1+h^2)$, respectively. For h negative, it is always true that $-(2h+7)/(1+h^2) > -7$, and so the optimal value is -7 and is always attained at $(0,1)$ for every negative h.

8. The dual problem is

$$\text{Minimize in } (y_1, y_2, y_3): \quad y_1 + y_2 + y_3$$

subject to

$$-y_1 + y_2 + y_3 \geq 1, \quad y_1 - y_2 + y_3 \geq 5, \quad y_1 + y_2 - y_3 \geq -3,$$

and $y_1, y_2, y_3 \geq 0$. This problem is equivalent to

$$(-)\text{Maximize in } (y_1, y_2, y_3): \quad -y_1 - y_2 - y_3$$

subject to

$$y_1 - y_2 - y_3 \leq -1, \quad -y_1 + y_2 - y_3 \leq -5, \quad -y_1 - y_2 + y_3 \leq 3,$$

and $y_1, y_2, y_3 \geq 0$. The minus sign in front of Maximize means that the value of the maximum is in need of a change of sign. If we now apply the duality process to this problem, we end up with

$$(-)\text{Minimize in } (z_1, z_2, z_3): \quad -z_1 - 5z_2 + 3z_3$$

under

$$z_1 - z_2 - z_3 \geq -1, \quad -z_1 + z_2 - z_3 \geq -1, \quad -z_1 - z_2 + z_3 \geq -1,$$

with $z_1, z_2, z_3 \geq 0$. It is clear that this problem is the initial one.

9. Due to the fact that all variables are negative, one needs first to change variables to $-x_1, -x_2, -x_3$. The optimal solution of the dual is then easily found graphically to be $(2, 6)$ with a minimum value -36. Duality conditions then lead to the conditions

$$3x_1 + x_2 = 3, \quad 2x_1 + x_3 = 5, \quad x_2 = 0,$$

with an optimal solution $(1, 0, 3)$ for the primal, and the same value -36 for the maximum.

10. The feasible set for this problem is unbounded in the first quadrant of the plane with three vertices $(8, 0)$, $(0, 3)$, and $(5/4, 9/8)$. Formally we also take into account the points at infinity $(0, +\infty)$, and $(+\infty, 0)$. A careful analysis of possibilities carry us to the following discussion:

   (a) The maximum is always $+\infty$ for every value of k.
   (b) For $k < 0$, the minimum is $-\infty$.
   (c) If $k = 0$, the minimum is zero, and is attained in every point of the vertical ray $[(0, 3), (0, +\infty)]$.
   (d) For $k \in (0, 2/3)$, the minimum is $3k$, and is attained at the unique vertex $(0, 3)$.
   (e) If $k = 2/3$, the minimum is 2, and taken on in the segment $[(0, 3), (5/4, 9/8)]$.
   (f) For $k \in (2/3, 6)$, the minimum is $5/4 + 9k/8$, and realized at $(5/4, 9/8)$.
   (g) If $k = 6$, the minimum is 8, and attained at the segment $[(5/4, 9/8), (8, 0)]$.
   (h) For $k > 6$, the minimum still is 8, but it is taken on in the ray $[(8, 0), (+\infty, 0)]$.

9.1.2 Practice Exercises

1. No comment.
2. No comment.
3. Use the dual. Find, with a detailed picture, that the solution for the dual is 240, though there is no unique optimal solution. This leads to an optimal solution for the primal $(1, 0, 0, 0)$ with a maximum of 240. Again, it is not unique.
4. If y_i are the dual variables, the information provided can be directly translated, through optimality and duality, into the facts

$$20y_1 + 10y_2 - y_3 = 22, \quad y_2 = 0, \quad 6y_1 + 2y_2 - 3y_3 = 6.$$

   The optimal solution of the dual is then $(10/9, 0, 2/9)$. Taking this information back into the duality conditions at optimality, we find that

$$2x_1 + 8x_2 + 6x_3 + x_5 = 20, \quad -x_1 + 5x_2 - 3x_3 - 8x_4 = -1, \quad x_1 = x_4 = 0.$$

   The optimal solution of the primal is $(0, 1, 2, 0)$.
5. Notice that the second constraint is redundant, as it is a consequence of the other two (by adding them up, and bearing in mind that $x_2 \geq 0$). This means that we can eliminate that constraint from the problem without changing it. Once this is clear, solve the dual graphically, and find the solution of the primal by duality. It is $(100, 100, 0)$.

6. Through the final formulation of Example 2.2 in Chap. 2, it is immediate to find the optimal solution (200, 40). To answer the following questions, one would, in principle, look at the dual problem in terms of the four tensions in the cables. Instead, in this case, it is quicker to argue as follows. For the optimal solution, the tensions are, calculated through the formulas in the explanation of the example,

$$T_A = T_B = 100, \quad T_C = 40, \quad T_D = 80.$$

Since these values do not get to the maximum allowable, except for T_B (there is a surplus of tension for the other three), we would conclude that this is the cable which would allow for an increase in the overall weight.

9.1.3 Case Studies

1. The primal and the dual are, respectively,

$$\text{Maximize in } \mathbf{x} = (x_1, x_2, x_3, x_4): \quad 20x_1 + 30x_2 + 30x_3 + 50x_4$$

subject to

$$x_1 + 2x_2 + 2x_3 + 4x_4 \leq 200, \quad 2x_1 + x_2 + 3x_3 + x_4 \leq 150, \quad x_1, x_2, x_3, x_4 \geq 0,$$

and

$$\text{Minimize in } \mathbf{y} = (y_1, y_2): \quad 200y_1 + 150y_2$$

under

$$y_1 + 2y_2 \geq 20, \quad 2y_1 + y_2 \geq 30, \quad 2y_1 + 3y_2 \geq 30, \quad 4y_1 + y_2 \geq 50, \quad y_1, y_2 \geq 0.$$

The solution of the dual is (40/3, 10/3) with a minimum value 9500/3. By using the duality relations, we find the optimal solution of the primal (100/3, 250/3, 0, 0). It is better to increase labor because the first component, the one corresponding to the labor constraint, is bigger.

2. The problem is easily set up as

$$\text{Maximize in } (x_1, x_2, x_3): \quad 3.35x_1 + 2x_2 + 0.5x_1$$

under

$$1.25x_1 + 0.75x_2 + 0.25x_3 \leq 8, \quad x_1, x_2, x_3 \geq 0.$$

This time the statement makes clear that variables x_i are supposed to be integer because a fraction of any element is worthless. If we ignore this feature, the dual problem leads easily to the optimal solution (6.4, 0, 0). Since the 0.4 of a table pays nothing, the clerk decides to make 6 tables, earning 20.1 euros, and he still has half an hour remaining, so that he can work on 2 stools, giving him an additional euro. Altogether, the optimal solution seems to be $(6, 0, 2)$ with a maximum salary of 21.1. This integer-like restriction on variables may be very hard to treat for more complicated problems.

3. The problem amounts to deciding the minimum of intensity of vigilance ρ over the park. Because ρ is linear, its minimum will be attained at one of the four vertices, and that will be the best point through which the thief may tray to escape. It is $(1, 3)$.

4. The problem can very clearly written in the form

$$\text{Minimize in } (x_A, x_B): \quad 30x_A + 35x_B$$

under

$$20x_A + 20x_B \geq 460, \quad 30x_A + 30x_B \geq 960, \quad 5x_A + 10x_B \geq 220, \quad x_A, x_B \geq 0.$$

As such, it is rapidly recognized as the dual of

$$\text{Maximize in } (y_H, y_N, y_P): \quad 460y_H + 960y_N + 220y_P$$

subject to

$$20y_H + 30y_N + 5y_P \leq 30, \quad 10y_H + 30y_N + 10y_P \leq 35, \quad y_H, y_N, y_P \geq 0.$$

However, because the primal is two-dimensional, it is more effective to solve it graphically to find that the optimal solution is $x_A = 14$, $x_B = 18$ with a minimal cost of 1050.

5. Though it is a three-dimensional problem, it can be solved graphically with a bit of care. Optimal solution: $(6, 4, 0)$, with a maximum profit of 1400.

6. Optimal solution: $(0, 32)$ with a maximum profit of 736.

7. If x_s and x_g indicate the number of scientific and graphic calculators, respectively, then we seek to
$$\text{Maximize in } (x_s, x_g): \quad -2x_s + 5x_g$$

under

$$x_s + x_g \geq 200, \quad 100 \leq x_s \leq 200, \quad 80 \leq x_g \leq 170, \quad x_s, x_g, \text{ integer.}$$

The optimal pair is $(100, 170)$ with a maximum profit 650.

8. If x and y stand for the number of cabinets of each corresponding type, we pretend to
$$\text{Maximize in } (x, y): \quad 8x + 12y$$

subject to

$$10x + 20y \le 140, \quad 6x + 8y \le 72, \quad x, y \ge 0, x, y, \text{ integer.}$$

If we neglect the integer constraint, the optimal pair maximizing storage volume is $(8, 3)$ which turns out to respect such constraint. Hence, it is the optimal solution with a maximum storage volume of 25 cubic feet.

9. (a) If (x_v, x_g, x_a) denote the quantities of the three instruments, respectively, the problem we face is to

$$\text{Maximize in } (x_v, x_g, x_a) : \quad 200x_v + 175x_g + 125x_a$$

subject to the constraints

$$2x_v + x_g + x_a \le 50, \quad 2x_v + x_g + 2x_a \le 60, \quad x_v + x_g + x_a \le 55,$$
$$x_v, x_g, x_a \ge 0, \quad x_v, x_g, x_a, \text{ integer.}$$

It is clear that

$$x_v + x_g + x_a \le 2x_v + x_g + x_a \le 50,$$

and so the third constraint is automatically satisfied once the first one is correct. That means that it can be eliminated without compromising the solution of the problem. The new version of the problem is then

$$\text{Maximize in } (x_v, x_g, x_a) : \quad 200x_v + 175x_g + 125x_a$$

subject to the constraints

$$2x_v + x_g + x_a \le 50, \quad 2x_v + x_g + 2x_a \le 60,$$
$$x_v, x_g, x_a \ge 0, \quad x_v, x_g, x_a, \text{ integer.}$$

(b) If (y_w, y_l) are the dual variables (prices of wood, and labor, respectively; price for metal would correspond to the constrained dropped off), the dual problem reads

$$\text{Minimize in } (y_w, y_l, y_m) : \quad 50y_w + 60y_l$$

under

$$2y_w + 2y_l \ge 200, \quad y_w + y_l \ge 175,$$
$$y_w + 2y_l \ge 125, \quad y_w, y_l \ge 0.$$

It is very easily seen graphically that the optimal solution is $(175, 0)$. By exploiting the duality relations, one finds that the optimal solution for the primal is $(0, 50, 0)$. Then the full solution of the dual for the first primal problem is $(175, 0, 0)$ with a vanishing optimal price for metal.

(c) The questions concerning extra units of wood, labor and metal refer precisely to the optimal solution of the dual problem: the company would pay 175 for each unit of wood, and nothing at all for labor or metal.

(d) The issue about the range of prices is related to the constraint, for dual prices, corresponding to violins

$$2y_w + 2y_l + y_m \geq 200.$$

While we change 200 to a number smaller that the value of the left-hand side for the optimal solution 350, the optimal production plan will not change. Thus violins could be sold at a maximum price of 350 euros without compromising the optimal production plan.

9.2   Chapter 3

9.2.1   Exercises to Support the Main Concepts

1. We are talking about the optimization problem

$$\text{Optimize in } \mathbf{x} \in \mathbb{R}^N : \quad \mathbf{x}^* \mathbf{A} \mathbf{x}$$

over unit vectors $|\mathbf{x}|^2 = 1$. Since this set is compact (bounded and closed), we can ensure that both the minimum and the maximum of the quadratic form are attained. KKT optimality conditions carry us to

$$2\mathbf{A}\mathbf{x} - 2\lambda\mathbf{x} = \mathbf{0}, \quad |\mathbf{x}|^2 = 1.$$

λ has to be an eigenvalue of $\mathbf{A}$, and by multiply by $\mathbf{x}$ itself, one sees that the two extrema correspond to the highest and smallest eigenvalues, respectively.

2. (a) The particular situation proposed can be written explicitly as

$$\text{Minimize in } (x, y) : \quad x^2 + y^2 \quad \text{under} \quad 2xy = 1.$$

By solving $y = 1/(2x)$ and replacing in the cost functional, it is very easy to find the optimal solution $\pm(1/\sqrt{2})(1, 1)$.

(b) If $\mathbf{x}$ is a non-null vector such that $\mathbf{x}^* \mathbf{Q} \mathbf{x} = 0$, take a sequence $\mathbf{x}_j$ with

$$\mathbf{x}_j \to \mathbf{x}, \quad |\mathbf{x}_j| = |\mathbf{x}|, \quad \mathbf{x}_j^* \mathbf{Q} \mathbf{x}_j = \alpha_j > 0.$$

It is then easy to check that $\tilde{\mathbf{x}}_j^* \mathbf{Q} \tilde{\mathbf{x}}_j = 1$, for

$$\tilde{\mathbf{x}}_j = \frac{1}{\sqrt{\alpha_j}}\mathbf{x}_j, \quad |\tilde{\mathbf{x}}_j| \to \infty (\alpha_j \to 0).$$

(c) If $\mathbf{A}$ is coercive, the minimum cannot be achieved at "infinity", and so it must be taken on some finite vector. KKT optimality conditions lead to the minimum value which is the minimum eigenvalue of $\mathbf{Q}^{-1}\mathbf{A}$.

3. The fact that the supremum of convex function is convex is standard, and easy to check directly through the definition. For the specific example, one obtain the convex function

$$\phi(\mathbf{x}) = \frac{1}{2\mathbf{a} \cdot \mathbf{x}}$$

if $\mathbf{a} \cdot \mathbf{x} > 0$. But $\phi(\mathbf{x}) = +\infty$ otherwise.

4. By looking at the problem

$$\text{Maximize in } \mathbf{y}: \quad \mathbf{x} \cdot \mathbf{y} \text{ under } |\mathbf{y}|^2 = 1,$$

is easy to establish the formula given. Conclude by the principle in the previous exercise that the resulting function of $\mathbf{x}$ is convex.

5. The basic properties of the convexification of a function are a consequence of the property that a supremum of convex functions is convex. For the given non-convex polynomial, treat first the case of $\psi_0(x) = (x^2 - 1)^2$, whose convexification is

$$\psi_0^\sharp(x) = 0 \text{ if } |x| \le 1, \quad \psi_0^\sharp(x) = \psi_0(x) \text{ if } |x| \ge 1.$$

Then argue that $\psi^\sharp(x) = x + \psi_0^\sharp(x)$.

6. This is pretty straightforward. The functions $\mathbf{x} \mapsto |\mathbf{x}|^p$ are strictly convex for all $p > 1$.

7. (a) No comment. The minimum is attained for $x = y = s/2$.

  (b) Take any two non-negative numbers x_0, y_0. Put $s = x_0 + y_0$. For this particular value of s, the first part of the problem means that

$$\frac{1}{2}\left[\left(\frac{x_0 + y_0}{2}\right)^n + \left(\frac{x_0 + y_0}{2}\right)^n\right] \le \frac{x_0^n + y_0^n}{2}.$$

Notice that the fact that the minimum is attained at $s/2$, means that

$$\left(\frac{s}{2}\right)^n \le \frac{x^n + y^n}{2}$$

for every couple (x, y), such that $x + y = s$. In particular, for (x_0, y_0) we should have

$$\left(\frac{x_0 + y_0}{2}\right)^n \le \frac{x_0^n + y_0^n}{2}.$$

This inequality can be proved by using the convexity of the function $x \mapsto x^n$ for $n \geq 1$, as well.

8. It is a matter of finding the optimal solution of the problem

$$\text{Minimize in } (x_1, x_2, \ldots, x_n): \quad \sum_i \frac{1}{2} x_i^2 \quad \text{under} \quad \sum_i x_i = a.$$

Conclude that the optimal partition is the equipartition $x_i = a/n$ for every i. Use then the optimality of this equipartition to argue as in the previous exercise to show the inequality.

9. Look at the problem

$$\text{Maximize in } (x_k)_{k=1,2,\ldots,n}: \quad \sum_k x_k y_k$$

under the condition

$$\sum_k x_k^p = b,$$

for $b > 0$, and numbers y_k. Find with care the optimal solution. Use again the maximality of the solution found to show the inequality.

10. Convince yourself first that if a real function f is convex, then

$$f\left(\sum_k t_k x_k\right) \leq \sum_k t_k f(x_k)$$

provided that $t_k \geq 0$, $\sum_k t_k = 1$. Then use this property for the convex function $f(x) = -\log(x)$, for $x > 0$. This function is convex because its second derivative is positive.

11. The critical values of x are $x = 0$ and $x = 2$. By differentiating once the differential equation, and replacing the derivative x' by $x^2(2 - x)$, it is easy to argue that x'' is non-negative around $x = 0$ so that x is convex in a vicinity of every t such that $x(t) = 2$, and these points are local minima, while it is non-positive around the value $x = 2$, and these are local maxima.

12. This is a simple example in which KKT conditions do not hold because the condition of the maximal rank in Theorem 3.1 is violated.

9.2.2 Practice Exercises

1. The first thing to realize is that, even though the feasible set could be unbounded, the coercivity of the objective function (it goes to infinity as vectors go to infinity in any way) implies that, at least the minimum is achieved in finite points which

must satisfy the KKT conditions. These reduce to

$$x_1 + \lambda x_2 + \lambda x_3 = \lambda x_1 + x_2 + \lambda x_3 = \lambda x_1 + \lambda x_2 + x_3 = 0,$$

together with $x_1 x_2 + x_1 x_3 + x_2 x_3 = 1$. The possible values of λ are those for which the determinant of

$$\begin{pmatrix} 1 & \lambda & \lambda \\ \lambda & 1 & \lambda \\ \lambda & \lambda & 1 \end{pmatrix}$$

vanishes. These are $\lambda = 1$, and $\lambda = -1/2$. The first possibility leads to vectors $x_1 + x_2 + x_3 = 0$, which together with the constraint imply that the value of the cost function must be -1 !!! To see this, put $0 = (x_1 + x_2 + x_3)^2$ and develop the square. Since the cost functional cannot take on negative values, this possibility is discarded. For $\lambda = -1/2$, we need $x_1 = x_2 = x_3 = \pm 1/\sqrt{3}$ with a cost value $1/2$. Since these are the only critical points, they must correspond to the minimum, the maximum has to be infinity, and, hence, the feasible set is unbounded.

2. We are talking about minimizing on the three coefficients of a parabola $ax^2 + bx + c$ the quadratic error at the four points given

$$(-1 - 16a - 4b - c)^2 + (a + b + c)^2 + (1 - a - b - c)^2 + (1 - a + b - c)^2$$

under no constraint for (a, b, c). Optimality conditions carry to the linear system

$$259a + 65b + 19c = -14, \quad 65a + 19b + 5c = -4, \quad 19a + 5b + 4c = 1,$$

whose solution is

$$a = -0.05, \quad b = -0.25, \quad c = 0.8.$$

It is instructive to draw the parabola $y = -0.05x^2 - 0.25x + 0.8$, and the four given points.

3. The feasible set is bounded and so the given function, which is smooth, attains both extreme values on the feasible set. Optimality conditions are written, through the fundamental theorem of Calculus to differentiate the integral,

$$\frac{-1}{1 + x^4} + 2\lambda x = 0, \quad \frac{1}{1 + y^4} + 2\lambda y = 0, \quad x^2 + y^2 = 1.$$

Because the polynomial $z \mapsto z + z^5$ is one-to-one, and odd, one can conclude that $y = -x$, necessarily, and then the two critical points are

$$\left(\frac{1}{\sqrt{2}}, -\frac{1}{\sqrt{2}} \right), \quad \left(-\frac{1}{\sqrt{2}}, \frac{1}{\sqrt{2}} \right).$$

The first corresponds to the minimum, and the second to the maximum.

4. If we interpret geometrically the function $x_1^2 + x_2^2$ as the square of the distance to the origen, then we are seeking the closest point of the feasible set to the origen. It is $(-1/5, 2/5)$. Same interpretation is valid for the second situation. This time the solution is $(1/3, 2/3)$.

5. It is easy to check that in the first situation the feasible set is empty. Concerning the second restriction set, note that we can use the linear constraint to neatly solve for one of the two variables in terms of the other: $x_1 = x_2 + 3$. Replacing this information back in the rest of the problem, we are left with an optimization problem in a single variable x_2 which is easily solved. The restriction set is $x_2 \in [-4 - 2\sqrt{2}, -4 + 2\sqrt{2}]$, and the objective function $3x_2^2 + 6x_2 + 9$. The minimum is attained at $x_2 = -4 + \sqrt{2}$, and the maximum at $-4 - 2\sqrt{2}$.

6. (a) The restriction set is a subset of the unit sphere, and so compact.
   (b) The one which is not convex is $x_1x_2 + x_1x_3 + x_2x_3$. Consider the section for $(x_1, x_2, x_3) = t(1, -1, 0)$.
   (c) Because of this lack of convexity, sufficiency of KKT is not guaranteed, nor uniqueness.
   (d) The careful analysis of KKT conditions and all possibilities carry us to the minimum point $(-1/\sqrt{3})(1, -1, 1)$ with a minimum value $-\sqrt{3}$.

7. (a) This is a particular situation of the first exercise with quadratic forms. The eigenvalues are -1 (twice) and 2.
   (b) For a situation with three independent variables and two constraints in the form of equalities, the part of the KKT conditions with derivatives can be expressed, without multipliers, as

$$\det(\nabla f, \nabla g_1, \nabla g_2) = 0.$$

   Argue why this is so, and apply to this particular situation. Upon factoring out this determinant, we find two possibilities: either $x = z$, or else $x + y + z = 0$. The first one leads to the maximum value $5/6$, while the second one takes us to the minimum $-1/2$.

8. Note first that the minimum is attained because the objective function is coercive. We can formally use KKT conditions. However, in this special situation where constraints simplify to $x, y \geq 0$, we can divide the discussion in four clear situations (which reflect the four situations with multipliers and KKT conditions), namely: $(0, 0)$; $(x, 0)$ with $x > 0$; $(0, y)$ with $y > 0$; and (x, y) with $x, y > 0$. Upon examining these four possibilities we conclude that the minimum is attained at $(1, 0)$ and its value is $-1/2$.

9. It is a matter of writing with some care the full set of KKT conditions, and check if all conditions concerning signs of the multipliers involved, vanishing products, etc., hold in each case. It is OK for the first point, but it is not so for the other two. The value of the multiplier associated with the second inequality constraint is inconsistent in two equations, while for the third, the second multiplier turns out to be negative.

10. The function f is strictly convex because it is the square of a non-singular linear (affine) transformation. The three problems admit a unique solution because all three can be written in the form of inequalities with convex functions, and an objective function which is strictly convex. These solutions are: $(0, 1)$ in the first situation (this is the absolute minimum of the objective function and it belongs to the first set); for the second example is impossible to find the solution with naked hands because the last step requires the solution of a cubic polynomial whose roots cannot be found easily in a exact, closed form; for the last case, and after a few computations, we find the optimal solution $(-3/5, 4/5)$.

11. The feasible set is definitely non-empty: the point $(1, 1)$, for instance, belongs to it. In addition, such non-empty admissible set is limited because it is a subset of the disc of radius 2 (centered at the origin). One can go through KKT optimality conditions, but in this case, by "completing" squares, we realize that the objective function can be recast as $(x_1 + x_2/2)^2 + (7/4)x_2^2$, and so the absolute minimum is zero, attained at the origin, provided that $(0, 0)$ is allowed by the constraints. It is so.

12. (a) Three simples examples of feasible points can be

$$(25, 100), \quad (10, 50), \quad (25, 10)$$

with respective costs $130/6, 62/6, 40/6$.

(b) The feasible region is unlimited. Note that the restrictions allow for arbitrary large and positive values of x_1 and x_2.

(c) The objective function grows indefinitely for those large and positive values of x_1 and x_2: the minimum is attained somewhere.

(d) All functions involved are linear (and hence convex) except for $25/x_1 + 100/x_2$. This is a strictly convex function (second derivatives strictly positive) over the feasible region. It is then easy to argue that KKT conditions furnish the unique optimal solution of the problem.

(e) For this example, it is not difficult to draw a picture of the feasible region, and the isocost lines. It is easy to argue that the point of minimum is $(10, 1000/95)$.

13. The restriction set is clearly compact, and so both extrema (of any continuous function) are attained. This is a very good practice exercise for optimality conditions. It requires a careful discussion of cases and subcases. After going through all possibilities, one finds that the candidates for the minimum are

$$(0, 0), (1, 0), (-2, 0), (3/2, \pm\sqrt{7}/2),$$
$$(1/2 + \sqrt{7}/2, -1/2 + \sqrt{7}/2), (1/2 - \sqrt{7}/2, -1/2 - \sqrt{7}/2),$$

while for the maximum are

$$(0, 0), (1, 0), (0, 2), (2, 0),$$
$$(1/2 + \sqrt{7}/2, -1/2 + \sqrt{7}/2), (1/2 - \sqrt{7}/2, -1/2 - \sqrt{7}/2).$$

The first two points occur in both lists because multipliers vanish, however the last two occur in both lists because they are the intersection points of the two constraints with equality, and it is rather tedious to decide on the sign of multipliers. Upon evaluating the objective function in all of these points we conclude that the minimum is $-14/3$ at $(-2, 0)$, while the maximum is 2 at $(0, 2)$.

9.2.3 Case Studies

1. For the first situation there are no restrictions to be respected, and the optimal location will minimize the sum of the square of the distances to the three points. Such an optimal location is $(130/3, 80/3)$. For the second part, the minimum of this sum of distances is to be found restricted to be located outside the park

$$(x_1 - 50)^2 + (x_2 - 90)^2 \geq 98.$$

The previous optimal location is not feasible. Optimality conditions need to be used. Two candidates are found in the points $(57, 13)$ and $(43, 27)$. Upon evaluating the objective function on both, the second one yields a smaller value.

2. (a) After redoing the computations of equilibria of forces and momenta, we find

$$\text{Maximize in } (x_1, x_2, x_3): \quad x_1 + x_2$$

subject to

$$x_1 x_3 \leq 1000, \quad 10x_1 - x_1 x_3 \leq 2000, \quad 25x_2 + 4x_1 x_3 \leq 5000, \quad 25x_2 + x_1 x_3 \leq 10000,$$
$$0 \leq x_1, \quad 0 \leq x_2, \quad 0 \leq x_3 \leq 10.$$

(b) Because the third variable always occurs in the product $x_1 x_3$, we put $z_3 = x_1 x_3$, and the problems changes to

$$\text{Maximize in } (x_1, x_2): \quad x_1 + x_2$$

subject to

$$10x_1 - 2000 \leq z_3 \leq 10x_1, \quad 25x_2 + 4z_3 \leq 5000, \quad 25x_2 + z_3 \leq 10000,$$
$$0 \leq x_1, \quad 0 \leq x_2, \quad 0 \leq z_3 \leq 1000,$$

which is a LP problem in the new set of variables.

(c) One can see graphically that the vertices of the feasible set correspond to either $z_3 = 0$, or $z_3 = 1000$. After examining these two situations, one finds the optimal triplet $(200, 200, 0)$ with maximum cost 400. This means that the lower load must be touching the left end-point of the bar (cable A).

3. It is elementary to find that the distance function x in terms of the angle u is

$$x = \frac{2v_0^2}{g}(-\sin^2 u + \sin u \cos u),$$

where g is gravity. It is therefore a matter of finding the maximum of this function in the interval $[0, \pi/2]$. Such a maximum corresponds to $u = \pi/8$.

4. This is a very simple problem in one variable. The optimal size of the small squares is $x = L/6$.

5. The Cobb–Douglas utility function is a very interesting example of a concave function (its negative is convex), which is the right concept for maximization problems, just as convexity is for minimum ones. If we let

$$\bar{x} = p_1\bar{x}_1 + p_2\bar{x}_2 + p_3\bar{x}_3$$

a number that indicates the total resources available for this consumer, then KKT optimality conditions lead to the optimal consumption vector

$$\bar{x} = \left(\frac{\alpha_1}{p_1}, \frac{\alpha_2}{p_2}, \frac{\alpha_3}{p_3}\right),$$

with maximum satisfaction

$$\bar{u} = \left(\frac{\bar{x}\alpha_1}{p_1}\right)^{\alpha_1} \left(\frac{\bar{x}\alpha_2}{p_2}\right)^{\alpha_2} \left(\frac{\bar{x}\alpha_3}{p_3}\right)^{\alpha_3}.$$

6. The problem to be solved is

$$\text{Maximize in } (x_1, x_2): \quad 7 - x_1^2 - 2x_2^2 + x_1x_2$$

under

$$2x_1 + x_2 \geq 1/2.$$

The condition $7 - x_1^2 - 2x_2^2 + x_1x_2 \geq 0$ need not to be explicitly imposed, as for the maximum of this function this constraint will automatically hold. Optimality conditions lead to the optimal point $(9/44, 1/11)$.

9.3 Chapter 4

9.3.1 Exercises to Support the Main Concepts

1. It is easy to find that
$$\mathbf{x}(y) = \left(-\frac{1}{2y}, \frac{1}{y}, \frac{1}{2y}\right),$$

and
$$G(y) = y \exp\left(\frac{3}{2y} - y\right).$$

There is a unique positive fixed point $y = \sqrt{3/2}$, and the corresponding $\mathbf{x}(y)$ is the optimal solution of the problem

$$\text{Minimize in } \mathbf{x}: \quad x_1 - 2x_2 - x_3 \quad \text{subject to } x_1^2 + x_2^2 + x_3^2 \leq 1.$$

2. (a) These two conditions are just the first and second, respectively, order optimality conditions for a local minimum of a real function of several variables. The whole point is to not get lost with computations involving the parameter y.
   (b) The previous first-order optimality condition determines implicitly $\mathbf{x}(y)$. Use implicit differentiation to find $\nabla \mathbf{x}(y)$, and use the preceding second-order optimality condition.
   (c) Note that
$$\nabla f(\mathbf{x}) + G(y)\nabla g(\mathbf{x}) = \mathbf{0},$$

   and differentiate this identity with respect to y.
3. This is just to reproduce, in this particular situation, the general algorithm described in the chapter.
4. This exercise amounts to applying Newton's method to approximate the unique fixed point of the function
$$G(y) = ye^{3/(2y)-y}$$

   for $y > 0$.
5. The conjugate gradient method for a quadratic functions, given that $\{\mathbf{e}_i\}$ is already an orthogonal basis, reads

   (a) Take an arbitrary $\mathbf{x}_0 \in \mathbb{R}^N$.
   (b) For each i, successively, put

$$\mathbf{x}_{i+1} = \mathbf{x}_i + t_i \mathbf{e}_i,$$

   where t_i is the minimum of the function of t

$$g(t) = q(\mathbf{x}_i + t\mathbf{e}_i).$$

Find an explicit formula for t_i. Check, by induction in i, that

$$\nabla q(\mathbf{x}_{i+1}) \cdot \mathbf{e}_k = 0, \quad k = 1, 2, \ldots, i$$

where $\cdot$ stands for the usual inner product in $\mathbb{R}^N$. In particular, for $i = N$,

$$\nabla q(\mathbf{x}_{N+1}) \cdot \mathbf{e}_k = 0, \quad k = 1, 2, \ldots, N,$$

$\nabla q(\mathbf{x}_{N+1}) = \mathbf{0}$, and $\mathbf{x}_{N+1}$ is the point of minimum of q.

6. This is an interesting example to see that the map of level curves can be pretty complicated, and the vector gradient a bit disconcerting. The negative of the gradient at $(0, 0)$ is $(2, 0)$ which is not particularly pointing towards the minimum.

7. The function whose minimum is desired is

$$E(\mathbf{x}) = \frac{1}{2}(x_1 - x_2 + 2x_3 - 2x_4 - 3)^2 + \frac{1}{2}(2x_1 + x_2 + 2x_3 - x_4 - 6)^2$$
$$+ \frac{1}{2}(x_1 + x_3 + x_3 + x_4 - 6)^2 + \frac{1}{2}(3x_1 - x_2 - 3x_3 + x_4)^2.$$

For a generic point $\mathbf{x}$, the value of t minimizing

$$E(\mathbf{x} - t\nabla E(\mathbf{x}))$$

can be written explicitly in terms of $\mathbf{x}$ because that particular value of t is the solution of a linear equation (with coefficients depending on $\mathbf{x}$). Try to compute several steps of the steepest descent method starting from $\mathbf{x}_0 = \mathbf{0}$.

9.3.2 Computer Exercises

1. This problem is a suggestion to write down your own subroutine to find the unrestricted minimum of a function of several variables. You can do so in C++, Fortran, Python, MatLab, etc. The result for the first one is $(0.997785, -0.929650)$. Note that it is easily seen that the point of minimum is $(1, -1)$. For the second one, the algorithm cannot find anything because the infimum is $-\infty$: the cost function tends to $-\infty$ as points move away to infinity along some direction, whereas for the third one, it is not so, and the point of minimum is, approximately, $(-1024.00, -1024.01, -0.643732)$ with a minimal cost -2046.31.

2. Writing a subroutine to accommodate constraints according to the algorithm described in this chapter is more challenging, but it is doable. The rest of the problems in this section are supposed to be solved with such a subroutine. Alternatively, you can use the subroutine provided by appropriate software packages.

3. It is worth to plot the feasible region. An approximation of the optimal solution
   is $(0.513296, -0.263273)$.
4. Starting out at $x_1 = x_2 = x_3 = x_4 = x_5 = 0$, one finds an approximation of the
   minimizer

$$(0.999761, 0.999956, 1.00005, 0.999942, 1.00000)$$

with a cost of $6.251161E - 08$. Admissibility criterium implies the (essentially)
non-positivity of the numbers

$$-1.735687E - 04, \quad 1.730919E - 04, \quad -1.158714E - 04,$$
$$1.158714E - 04, \quad -5.912781E - 05, \quad 5.912781E - 05,$$

while optimality is meant by the nearly vanishing of

$$-1.731555E - 04, \quad 1.734418E - 04, \quad -1.155524E - 04,$$
$$1.161401E - 04, \quad -5.899371E - 05, \quad 5.924471E - 05.$$

5. Starting from $x_1 = x_2 = x_3 = 0$, and auxiliary variables $y_1 = y_2 = 1$, the
   optimal solution is
$$(1.14520, 1.25657, 1.33411).$$

   Recall that an equality constraints is equivalent to two inequalities.
6. The approximation found is

$$(0.577376, 0.577462, 0.577191)$$

using the point $(1, -1, 1)$ as a initialization with auxiliary variables y set to unity.
7. The polynomial turns out to be non-negative. Indeed, it is easy to check (though
   not so easy to find) that

$$P(x_1, x_2) = \frac{1}{2}(2x_1^2 - 3x_2^2 + x_1 x_2)^2 + \frac{1}{2}(x_2^2 + 3x_1 x_2)^2.$$

The vanishing minimum is attained at $(0, 0)$.
8. Both polynomials are non-negative. However they cannot be rewritten as a sum
   of squares. The first case can be recast as

   Minimize in (s, t) : $st(s + t - 1)(+1)$ under $s, t \geq 0, s + t \leq 1.$

The minimum is taken on at $(1/3, 1/3)$ with a minimum value $1/27$. The minimum for the second one clearly vanishes at $(0, 0, 0)$.
9. The minimum should be non-negative. In fact, it is easy to check that the polynomial can be rewritten in the form

$$x_2^2 x_1 + x_1^2 x_2 + (x_1 - x_2)^2 (x_1 + x_2 - 1),$$

which shows that, in the constraint set, the polynomial is non-negative. This is a very nice example to realize of the difficulties in approximating (local) minima when a non-convex function, like the cubic polynomial in this case, is involved. Yet a suitable change of variables may allow us to find the global minimum. Put $x + y = z$, $z \geq 1$, and use this identity to eliminate y from the formulation. You will find that the resulting cost functional is quadratic in x (with coefficients depending on z). Its minimum has to be found in the interval $[0, z]$, $z \geq 1$. This is easily done: the minimum is attained at $z/2$, and its value is $z^3/4$ provided $z \geq 4/3$, while it is $z^3 - z^2$ for $1 \leq z \leq 4z/3$, and is taken on at $x = 0, z$. Altogether, the vanishing absolute minimum corresponds to the two (symmetric) points $(1, 0)$, $(0, 1)$.

10. There should not be any particular difficulty in finding the smallest eigenvalue of the matrix suggested: -1.09990 is the eigenvalue, and

$$(5.55E - 03, 0.706756, -0.707356),$$

the corresponding eigenvector.

9.3.3 Case Studies

1. It is easy to set up the general format for a basic transportation problem, namely,

$$\text{Minimize in } \mathbf{x} = (x_{ij}) : \quad \sum_i \sum_j c_{ij} x_{ij}$$

under

$$\sum_i x_{ij} \geq v_j \text{ for every } j,$$

$$\sum_j x_{ij} \leq u_i \text{ for every } i,$$

$$x_{ij} \geq 0 \text{ for every } i, j.$$

In the particular situation proposed, we would have nine variables x_i, $i = 1, 2, \ldots, 8, 9$, identified with the entries of the matrix row after row, so that the cost is

$$x_1 + 2x_2 + 3x_3 + 2x_4 + x_5 + 2x_6 + 3x_7 + 2x_8 + x_9$$

and we would have six linear constraints

$$x_1 + x_2 + x_3 \leq 2, \quad x_4 + x_5 + x_6 \leq 3, \quad x_7 + x_8 + x_9 \leq 4,$$
$$x_1 + x_4 + x_7 \geq 5, \quad x_2 + x_5 + x_8 \geq 2, \quad x_3 + x_6 + x_9 \geq 2,$$

in addition to having all variables non-negative. The optimal solution that is obtained from the starting point $x_i = 0$ for all i (an auxiliary variables $y = 1$), is

$$(1.99978, 1.122965E - 04, 8.103371E - 05, 1.91547,$$
$$1.08475, 1.416386E - 04, 1.08467, 0.915258, 1.99993)$$

with a minimal cost 14.0007, from which we can infer the (one) exact solution

$$(2, 0, 0, 2, 1, 0, 1, 1, 2)$$

with cost 14. It is not however the unique solution, because

$$(2, 0, 0, 1, 2, 0, 2, 0, 2)$$

is another one (with the same cost).

2. The diet problem is also a classical example in linear programming. If x_i are the content of food i in a unitary diet, then we seek to

$$\text{Minimize in } \mathbf{x} = (x_i) : \quad \sum_i c_i x_i$$

subject to

$$\sum_j a_{ij} x_i \geq b_j \text{ for every nutrient } j, \quad x_i \geq 0.$$

a_{ij} is the content of nutrient j in food i, and b_j is the minimum amount required of each nutrient j. In the particular situation described in this problem we would have to

$$\text{Minimize in } (x_1, x_2, x_3, x_4, x_5) : \quad x_1 + 0.5x_2 + 2x_3 + 1.2x_4 + 3x_5$$

subject to

$$78.6x_1 + 70.1x_2 + 80.1x_3 + 67.2x_4 + 77.0x_5 \geq 74.2,$$
$$6.50x_1 + 9.40x_2 + 8.80x_3 + 13.7x_4 + 30.4x_5 \geq 14.7,$$
$$0.02x_1 + 0.09x_2 + 0.03x_3 + 0.14x_4 + 0.41x_5 \geq 0.14,$$
$$0.27x_1 + 0.34x_2 + 0.30x_3 + 1.29x_4 + 0.86x_5 \geq 0.55,$$
$$x_1, x_2, x_3, x_4, x_5 \geq 0.$$

By using our algorithm with exponential barriers, and using as initialization $x_i = 1$ for all i, and $y_j = 1.$ for all j, the solution found is, approximately,

$$(2.521589E-04, 1.53086, 6.266541E-05, 2.260657E-02, 6.016344E-05)$$

with a minimal cost 0.793118. Parameters indicating admissibility are

$$- 34.6622, \quad -3.847599E - 03, \quad -9.742330E - 04, \quad 2.053063E - 04,$$
$$- 2.521589E - 04, \quad -1.53086, \quad -6.266541E - 05,$$
$$- 2.260657E - 02, \quad -6.016344E - 05,$$

which are nearly non-positive, while certificate of convergence to the true solution is ensured because all numbers

$$- 2.282715E - 42, \quad -1.216474E - 04, \quad -1.065329E - 05, \quad 1.219519E - 04,$$
$$- 1.603842E - 04, \quad -1.046452E - 05, \quad -9.689064E - 05,$$
$$- 1.026057E - 05, \quad -9.187412E - 05,$$

are almost zero.

3. The various restrictions given for the portfolio situation lead to be concerned with the problem

$$\text{Maximize in } \mathbf{x} = (x_i) : \quad \sum_i d_i(b_i + x_i)$$

subject to

$$b_i + x_i \geq 0 \text{ for all } i, \quad r \sum_i v_i(b_i + x_i) \leq v_j(b_j + x_j) \text{ for all } j,$$

$$\sum_i v_i x_i = 0, \quad \sum_i w_i(b_i + x_i) \geq (1+s) \sum_i v_i b_i.$$

Specific data carry us to be concerned about

$$\text{Minimize in } (x_1, x_2, x_3) : \quad -3x_2 - 5x_3(-475)$$

subject to the conditions

$$- x_1 - 75 \leq 0, \quad -x_2 - 100 \leq 0, \quad -x_3 - 35 \leq 0,$$
$$- 3x_1 + x_2 + 5x_3 + 50 \leq 0, \quad x_1 - 3x_2 + 5x_3 - 50 \leq 0,$$
$$x_1 + x_2 - 15x_3 - 350 \leq 0,$$
$$x_1 + x_2 + 5x_3 \leq 0, \quad -x_1 - x_2 - 5x_3 \leq 0,$$
$$- 10 - 18x_1 - 23x_2 - 102x_3 \leq 0.$$

Many of these linear expressions involve evaluations of big numbers. If we bear in mind that we are using exponential barriers, it is important to rescale variables so that feasible values are of the order of unity. If we put $X_i = x_i/100$, in these new variables the problem would read

$$\text{Minimize in } (X_1, X_2, X_3) : \quad -3X_2 - 5X_3(-4.75)$$

subject to the conditions

$$- X_1 - 0.75 \le 0, \quad -X_2 - 1 \le 0, \quad -X_3 - 0.35 \le 0,$$
$$- 3X_1 + X_2 + 5X_3 + 0.5 \le 0, \quad X_1 - 3X_2 + 5X_3 - 0.5 \le 0,$$
$$X_1 + X_2 - 15X_3 - 3.5 \le 0,$$
$$X_1 + X_2 + 5X_3 \le 0, \quad -X_1 - X_2 - 5X_3 \le 0,$$
$$- 0.001 - 0.18X_1 - 0.23X_2 - 1.02X_3 \le 0.$$

We use initialization $X_i = 0$ for all i, and auxiliary variables for constraints

$$y(1) = y(2) = y(3) = y(7) = y(8) = y(9) = 0.5, \quad y(4) = y(5) = y(6) = 20.$$

The optimal solution turns out to be

$$(0.125003, 0.749935, -0.175005)$$

complying with all certificates of admissibility and convergence. We therefore infer that the optimal solution for the original portfolio problem is $(12.5, 75, -17.5)$ with a maximum benefit in dividends of 612.5.

9.4 Chapter 5

9.4.1 Exercises to Support the Main Concepts

1. (a) Note that $u_t(x) = x^t$ is feasible because $u_t(0, 1) = 0, 1$. Its derivative is $u'_t(x) = tx^{t-1}$, and so

$$E(u_t) = \frac{1}{2} \int_0^1 t^2 x^{2(t-1)} \, dx.$$

If $t \le 1/2$, the resulting integral is divergent. Its value is $+\infty$. Whereas for $t \ge 1/2$, the value of the integral is $t^2/(2t - 1)$. The minimum of this function on that range of t corresponds to $t = 1$, and the minimum value of the integral is 1.

(b) Once the integral is written, and one term eliminated because of periodicity, it is easy to see that $n = 0$ is the best.

(c) Put $w(x) = x(1 - x)v(x)$. Show that

$$\int_0^1 [(1 + w'(x))^2 - 1] \, dx = \int_0^1 w'(x)(2 + w'(x)) \, dx = \int_0^1 w'(x)^2 \, dx \geq 0.$$

Thus $v \equiv 0$ is the best possibility.

2 (a) It is clear, because there are no derivatives involved in the functional nor any constraints at end-points of the interval, that the minimizer will take place minimizing the integrand for every x in the interval of integration $u(x) = x/2$.

(b) The function $u(x) = x/2$ respects the value at the left end-point, but not at the right end-point. That means that the infimum is not attained as we can maintain the value $x/2$ most of the time in the interval $[0, 1]$, except for a small subinterval near 1, where a rapid transition takes us to respect $u(1) = 1$.

(c) The E-L equation for the problem is

$$-u''(x) + 2u(x) - x = 0 \text{ in } (0, 1), \quad u(0) = 0, u(1) = 1.$$

The solution is

$$u(x) = \frac{1}{2} \frac{e^{\sqrt{2}x} - e^{-\sqrt{2}x}}{e^{\sqrt{2}} - e^{-\sqrt{2}}} + \frac{x}{2}.$$

(d) If we introduce that small parameter ε, the minimizer becomes

$$u_\varepsilon(x) = \frac{1}{2} \frac{e^{\sqrt{\frac{2}{\varepsilon}}x} - e^{-\sqrt{\frac{2}{\varepsilon}}x}}{e^{\sqrt{\frac{2}{\varepsilon}}} - e^{-\sqrt{\frac{2}{\varepsilon}}}} + \frac{x}{2}.$$

3. Consider the function of ε

$$\int_0^L F(x, u(x) + \varepsilon U(x), u'(x) + \varepsilon U'(x), u''(x) + \varepsilon U''(x)) \, dx$$

for $U(x)$ such that $U(0) = U(1) = U'(0) = U'(1) = 0$. Differentiate with respect to ε, and perform the necessary integrations by parts to find that

$$\frac{d^2}{dx^2} F_p(x, u(x), u'(x), u''(x)) - \frac{d}{dx} F_z(x, u(x), u'(x), u''(x))$$
$$+ F_u(x, u(x), u'(x), u''(x)) = 0,$$

where $F = F(x, u, z, p)$. For the particular case proposed, the equation is $u'''' + u = 0$, and the minimizer

$$u(x) = e^{(\sqrt{2}/2)x} \cos\left(\frac{\sqrt{2}}{2}x\right).$$

4. The usual procedure of using "variations" with $U(0) = U(1) = 0$ leads to the E-L equation as usual. Once this is known, and bearing in mind that in fact $U(0) = U(1)$ could be any arbitrary value, conclude that

$$F_z(0, u(0), u'(0)) = F_z(1, u(1), u'(1)), \quad u(0) = u(1).$$

For the particular proposed example, the minimizer is the solution of the problem

$$-u''(x) + u(x) - x = 0 \text{ in } (0, 1), \quad u(0) = u(1), u'(0) = u'(1).$$

The solution is

$$u(x) = x + \frac{1}{2(1 - e)}e^x + \frac{1}{2(1 - e^{-1})}e^{-x}.$$

5. Differentiate with respect to a given Fourier coefficient a_k, integrate by parts, and conclude that a certain expression (the E-L equation) is such that the integral against all functions $\cos(2\pi kx)$ vanishes. Conclude that the E-L equation must then vanish.
6. The E-L equations in the corresponding chapter have been written for a vector problem, so that this exercise asks to write explicitly that system for a two dimensional problem with components $u(x)$ and $v(x)$. Conclude that the solution for each component is the linear function determined by the end-point conditions.
7. Argue that

$$v(t, y) = \min_{w}\{\min_{u}\left\{\int_t^s F(x, u(x), u'(x))\,dx : u(t) = y, u(s) = w\right\} + v(s, w)\},$$

if $t < s < 1$. Reorganize this equality by introducing $v(t, y)$ into the outer minimum, and put $w = y + z(s - t)$ to conclude.
8. The whole point is to check that the second-order E-L system written for the lagrangian of the problem

$$-\frac{d}{dt}L_z(\mathbf{q}(t), \mathbf{q}'(t)) + L_q(\mathbf{q}(t), \mathbf{q}'(t)) = \mathbf{0},$$

can be interpreted in terms of the Hamiltonian through the first-order Hamilton's equations. The definition of the Hamiltonian leads to the relations

$$H(\mathbf{q}, \mathbf{p}) = \mathbf{z}(\mathbf{q}, \mathbf{p}) \cdot \mathbf{p} - L(\mathbf{q}, \mathbf{z}(\mathbf{q}, \mathbf{p})), \quad \mathbf{p} - L_z(\mathbf{q}, \mathbf{z}(\mathbf{q}, \mathbf{p})),$$

where this $\mathbf{z}$ is the optimal vector $\mathbf{z}$ (depending on $\mathbf{q}$ and $\mathbf{p}$) in the definition of the hamiltonian. Then it is easy to check that

$$L_{\mathbf{z}}(\mathbf{q}, \mathbf{z}) = \mathbf{p}, \quad \mathbf{z} = H_{\mathbf{p}}(\mathbf{q}, \mathbf{p}), \quad H_{\mathbf{q}}(\mathbf{q}, \mathbf{p}) = -L_{\mathbf{q}}(\mathbf{q}, \mathbf{z}),$$

If we put $\mathbf{q}' = \mathbf{z}$, through these identities it is immediate to check the equivalence between the E-L system and Hamilton's equations.

9. Either check that the solution of the E-L equation is incompatible with the end-point conditions, or use the identity $tu'(t) = (tu(t))' - u(t)$ to eliminate derivatives from the functional.

10. Just realize that

$$[u(t) f(u(t))]' = u'(t) f(u(t)) + u(t) f'(u(t)) u'(t).$$

Likewise, the integrand for the second functional is $[F(u(t), u)]'$. The E-L equation for these integrands is an identity independently of the function u.

11. For a general functional of the form

$$\int_\Omega F(u(\mathbf{x}), \nabla u(\mathbf{x})) \, d\mathbf{x}, \quad F = F(u, \mathbf{U}),$$

the E-L equation is a partial differential equation of the form

$$-\operatorname{div} F_{\mathbf{U}}(u, \nabla u) + F_u(u, \nabla u) = 0 \text{ in } \Omega.$$

This can be formally shown by using the method of variations. In the particular case of the Dirichlet integral, the equation turns out to be $-\Delta u = 0$, and so minimizers are harmonic functions. For the vector example with the determinant, it is easy to check that the E-L equation for each component evaporates because of the equality of the mixed partial derivatives of a smooth function: you get nothing $0 = 0$. This can be interpreted by saying that every function is a minimizer as long as it complies with the given boundary condition. This is exactly what it means being a null-Lagrangian: the functional only depends on the boundary datum, and not on what the values are inside Ω. This can also be seen recalling the formula of the change of variables for two dimensional integrals.

9.4.2 Practice Exercises

1. No comment.
2. The minimizer is $y(x) = e^x - e^{-x}$.
3. The minimizer is $u(x) = (e^x - e^{-x})/(e - e^{-1})$. Notice that the term $u(x)u'(x)$ is a null-lagrangian, so it does not add anything to the E-L equation.

4. The transversality condition has to be utilized at both end-points. The minimizer is $y(x) = (1/3)e^x - (2/3)e^{-x}$, and the minimum value of the integral is $\log 2 - 2/9$.

5. With a bit of patience, it is straightforward to find

$$y_s(x) = \frac{2s}{3}(e^x - e^{-x}), \quad f(s) = \frac{15}{9}s^2 + s.$$

The minimum of this function takes place at $s = -3/10$, and so the optimal function will be

$$y(x) = -\frac{1}{5}(e^x - e^{-x}).$$

This same optimal y is obtained by using the transversality condition $2y' + 1 = 0$ at the right end-point $x = \log 2$.

6. A multiplier z (an unknown number) is introduced to modify the integrand from $y'(x)^2$ to $y'(x)^2 + zy(x)$. After using the E-L equation for this new integrand, and demanding the three conditions

$$y(0) = y(1) = 0, \quad \int_0^1 y(x)\,dx = 1,$$

it is elementary to find the optimal $y(x) = 6x - 6x^2$. The other method suggested leads to $y(x) = z'(x)$, $y'(x) = z''(x)$, and so the integrand to be minimized is $z''(x)^2$ under the four end-point conditions $z(0) = z'(0) = z'(1) = 0$, $z(1) = 1$. The optimal function is $z(x) = x^2(3 - 2x)$, and its derivative $z'(x) = 6x - 6x^2$ is the optimal $y(x)$.

7. The transit functional is

$$\int_0^1 \frac{\sqrt{1 + u'(x)^2}}{x}\,dx.$$

The corresponding E-L equation reduces to

$$\frac{u'(x)}{x\sqrt{1 + u'(x)^2}} = c$$

for some constant c. Various manipulations, and a direct integration carry us to

$$u(x) = -\frac{1}{c}\sqrt{1 - c^2x^2} + d.$$

Conditions at the end-points translate into

$$u(x) = 1 - \sqrt{1 - x^2}.$$

What curve is this?

8. The function $\phi(u, z) = (z^2 + uz)^2$ is not convex in z. So the conditions to guarantee existence of solutions do not hold. Nevertheless the function $u(x) \equiv 0$ is trivially a minimizer. Once we know that zero is the minimum, it is easy to argue that the trivial function is the only minimizer. Note that the only u with $u'(x)^2 + u(x)u'(x) = 0$ under $u(1) = 0$ is the trivial one.

9. The function $\phi(x, z) = z^2 + xz$ is strictly convex regardless of the value of x. Therefore there is a unique minimizers under whatever end-point conditions. The minimizer complying with the only condition $y(0) = 1$ is $y(x) = -(1/4)x^2 + 1$.

10. It is pretty clear that the unique minimizer will be the solution of the Cauchy problem given, whose solution is easily found.

11. The function $\phi(z) = z^2 + 1/z$ is strictly convex if $z > 0$. Therefore the unique minimizer of the problem is the linear function determined by the end-point values.

12. The functional whose E-L equation is given is

$$\int_0^1 [\frac{1}{2}u'(t)^2 + G(u(t))] \, dt$$

where $G(u)$ is any primitive of $g(u)$: $G'(u) = g(u)$. If g is non-decreasing so that $g' = G'' \geq 0$, G will be convex, and there will be a unique solution.

13. If $y(1)$ is free, the minimizer has already been found in a previous exercise: $y(x) = -x^2/4 + 1$, whose value at $x = 1$ is $3/4$. Because this value is non-feasible for the other two situations, optimal solutions will be the solutions of the E-L equation with $y(1) = 1$, and $y(1) = 1/2$, respectively, which are the closest permitted valued to $3/4$ for each case.

14. The feasible region for the mathematical program is convex but non-bounded. The minimum is, however, attained because the objective function is coercive, and strictly convex. The point of minimum is $(3/2, 0)$. The E-L equation for the variational problem, under an integral constraint, is $u''(t) = \lambda$ for a certain multiplier λ. Thus

$$u(t) = \frac{\lambda}{2}t^2 + w_1 t + w_2$$

for another two unknown constants w_1 and w_2. It is a matter of writing all of the ingredients of the problem in terms of these three constants (λ, w_1, w_2), by using the previous $u(t)$, to find

$$\text{Minimize in } (\lambda, w_1, w_2) : \quad \frac{\lambda^2}{6} + \frac{\lambda w_1}{2} + \frac{w_1^2}{2}$$

subject to

$$0 \leq w_2 \leq 1, \quad \frac{\lambda}{2} + w_1 + w_2 = 1, \quad \frac{\lambda}{6} + \frac{w_1}{2} + w_2 \leq \frac{1}{2}.$$

The equality constraint can be used to eliminate w_2, and putting $x = \lambda$, $y = w_1$, we are led to the mathematical program in the statement. The optimal solution found $x = 3/2$, $y = 0$, carries us to the optimal

$$u(t) = \frac{3}{4}t^2 + \frac{1}{4}.$$

9.4.3 Case Studies

1. If P and Q are those two points $|P|^2 = |Q|^2 = 1$, and $\gamma(t) : [0, 1] \to \mathbf{S}$ is an arbitrary path on the sphere joining them

$$\gamma(0) = P, \gamma(1) = Q, \quad |\gamma(t)|^2 = 1 \text{ for all } t \in [0, 1],$$

the functional to be minimized under these constraints is

$$L(\gamma) = \int_0^1 |\gamma'(t)| \, dt.$$

The E-L system (three equations) yields $\gamma'(t) = \mathbf{v}$ a constant vector, that has to be orthogonal to $\gamma(t)$. That means shortest paths must be contained in a plane through the origen, and so it is an arc of a maximum circle (a meridian) through the two points.

2. For the first part, use a multiplier $1/z$ to account for the integral constraint

$$\int_0^1 \sqrt{1 + u'(t)^2} \, dt = B.$$

Write the E-L equation for $u(t) + (1/z)\sqrt{1 + u'(t)^2}$, and, after a few algebraic manipulations, conclude that the optimal u is of the form

$$u(t) = \sqrt{1 - (zt + w)^2} + r,$$

for constant z, w, and r. These are to be found through the three constraints

$$u(0) = 0, u(1) = 1, \int_0^1 \sqrt{1 + u'(t)^2} \, dt = B.$$

Beyond this particular adjustment, the interesting point is to conclude that the optimal curve is an arc of a circle. Note that

$$(u(t) - r)^2 + (zt + w)^2 = 1.$$

For the second part, the objective functional and the integral constraint change roles. Computations are pretty similar. The optimal curve is again an arc of a circle.

3. The problem for the geodesics in that surface would be

$$\text{Minimize in } \gamma(t) : \quad \int_0^1 |\gamma'(t)| \, dt$$

subject to $\gamma_2(t) = cosh(\gamma_1(t))$, $P = \gamma(0)$, $Q = \gamma(1)$. For the two points $P = (0, 1, 0)$, $Q = (1, cosh\, 1, 1)$, we can assume that $\gamma_1(t) = t$, and $\gamma_2(t) = cosh(t)$, and the geodesic will be the minimizer of the functional

$$\int_0^1 \sqrt{cosh^2(t) + \gamma_3'(t)^2} \, dt$$

under

$$\gamma_3(0) = 0, \quad \gamma_3(1) = 1.$$

This is a variational problem for the single unknown function $\gamma_3(t)$. Using in a fundamental way that

$$cosh'x = sinh\,x, \quad sinh'x = cosh\,x \quad 1 + sinh^2x = cosh^2x$$

the associated E-L equation leads to

$$\gamma_3(t) = \frac{1}{sinh\,1} sinh\,t.$$

4. Feasible profiles $y(x)$ must respect the end-point conditions $y(0) = y(1) = 0$. In particular, $y(x) = -x(1-x)$ does. The optimal profile will be the one minimizing energy under the end-point conditions

$$-y''(x) + g(x) = 0 \text{ in } (0, 1), \quad y(0) = y(1) = 0.$$

The (unique) solution is

$$y(x) = \frac{x^4}{12} - \frac{x^3}{6} + \frac{x}{12}.$$

5. Evidently, at the ceiling we should ask for $u(0) = 0$, but at the free end-point nothing should be imposed because that condition ought to be the result of minimizing energy: the natural or transversality condition. The two given possibilities are, then, feasible. Their respective energies are easily calculated $-17W^2/96$, $-5W^2/16$, respectively. To find the configuration of minimal energy, we must solve the problem

$$-[(1+x)u'(x) - W]' = 0 \text{ in } (0, 1), \quad u(0) = 0, 2u'(1) - W = 0.$$

Note the natural boundary condition at $x = 1$. The unique solution is

$$u(x) = W \log(1 + x)$$

with energy $- \log 2W^2/2$.

6. A multiplier z needs to be introduced to account for the constraint on the area, namely, we ought to care about the E-L system (because we have two unknowns) for the integrand

$$F(u, v, p, q) = \sqrt{p^2 + q^2} + z(uq - vp).$$

The two equations are

$$-\frac{d}{dt}\frac{\partial F}{\partial p} + \frac{\partial F}{\partial u} = 0, \quad -\frac{d}{dt}\frac{\partial F}{\partial q} + \frac{\partial F}{\partial v} = 0,$$

which, in our particular case, look like

$$-\frac{d}{dt}\left(\frac{u'(t)}{\sqrt{u'(t)^2 + v'(t)^2}} - zv(t)\right) + zv'(t) = 0,$$

$$-\frac{d}{dt}\left(\frac{v'(t)}{\sqrt{u'(t)^2 + v'(t)^2}} + zu(t)\right) - zu'(t) = 0.$$

A first integration leads to

$$\frac{u'(t)}{\sqrt{u'(t)^2 + v'(t)^2}} - 2zv(t) = c, \quad \frac{v'(t)}{\sqrt{u'(t)^2 + v'(t)^2}} + 2zu(t) = d.$$

Appropriate manipulation of these two equations leads to

$$(u(t) - C)^2 + (v(t) - D)^2 = r^2$$

for suitable constants C, D, and r. This shows that indeed the optimal curve is a circle.

If σ is an arbitrary closed curve of the plane enclosing an area $A = A(\sigma) > 0$, its length $L(\sigma)$ is greater than that of the optimal profile $\tilde{\sigma}$ enclosing that same area $A = A(\tilde{\sigma})$. But we have argued that $\tilde{\sigma}$ is a circle, and so $L(\tilde{\sigma})^2 = 4\pi A(\tilde{\sigma})$. Then

$$L(\sigma)^2 \geq L(\tilde{\sigma})^2 = 4\pi A(\tilde{\sigma}) = 4\pi A = 4\pi A(\sigma).$$

7. This is the classic problem of optimal Newton's solid of revolution. It is very easy to check that the drag for the line and the circle is 1/2. The integrand, as a function of y', is not convex. As a matter of fact, the central part around the origin is concave, and then it has two symmetric inflection points. The E-L equation is easy to write, but impossible to solve by hand. As a matter of fact, the optimal profile consists of two distinct parts: one inner flat part $[0, 0.351]$, and an increasing part up to 1. If the functional changes to the other proposal, convexity properties are also compromised so that solutions of the E-L equation may not be minimizers. Yet more steps can be taken toward the solution of it.

8. The associated hamiltonian is calculated through the formula

$$H(\mathbf{p}, \mathbf{q}) = \max_{\mathbf{z} \in \mathbb{R}^N} \left(\mathbf{p} \cdot \mathbf{z} - \frac{1}{2}m|\mathbf{z}|^2 + \frac{1}{2}k|\mathbf{q}|^2 \right).$$

It is a simple exercise to find that

$$H(\mathbf{p}, \mathbf{q}) = \frac{1}{2m}|\mathbf{p}|^2 + \frac{1}{2}k|\mathbf{q}|^2,$$

the total energy of the system, which is a preserved quantity of the motion. The equations of motion are easily written. It is interesting to realize that despite the fact that the Lagrangian is convex on the derivative $\mathbf{x}'$, it is not so with respect to $\mathbf{x}$, because of the minus sign in front of the square. As a matter of fact there cannot be minimizers because the action integral is not bounded from below: for a constant path $\mathbf{x} \equiv \mathbf{u}$, the action is $-(1/2)k|\mathbf{u}|^2$. Hence, motion takes places along equilibria (just solutions of E-L) of the action integral but not minimizers.

9. In the same spirit of the previous problem, write the equation of motion as

$$-\frac{1}{2}m\mathbf{x}'' - \nabla V(\mathbf{x}) = 0.$$

Then

$$V(\mathbf{x}) + \frac{m}{4}|\mathbf{x}'|^2 = c = V(\mathbf{x}_0) + \frac{m}{4}|\mathbf{x}_0'|^2,$$

if $(\mathbf{x}_0, \mathbf{x}_0')$ is the state at the starting time. Then

$$V(\mathbf{x}) \leq V(\mathbf{x}_0) + \frac{m}{4}|\mathbf{x}_0'|^2,$$

is a bound for the potential throughout the motion.

9.5 Chapter 6

9.5.1 *Exercises to Support the Main Concepts*

1. If one neglects the pointwise constraint $u(x) \leq 1$, the optimal solution turns out to be

$$u(x) = 10 + 10\frac{e^{-1} - 1}{e - e^{-1}}e^x + 10\frac{1 - e}{e - e^{-1}}e^{-x}.$$

This function is greater than unity in an inner subinterval around $1/2$. Therefore the pointwise constraint $u(x) \leq 1$ needs to be active somewhere if enforced. The new functional with the exponential perturbation is both strictly convex and coercive for fixed $v(x) > 0$. The solution is unique for each such $v(x)$. The optimal solution for the constrained variational problems is plotted in Fig. 9.1.

2. It is clear that the inequality has to be active somewhere because the solution of the variational problem without enforcing this inequality is the trivial function. See a plot of the optimal solution together with the "obstacle" (in black) in Fig. 9.2.

3. If you write

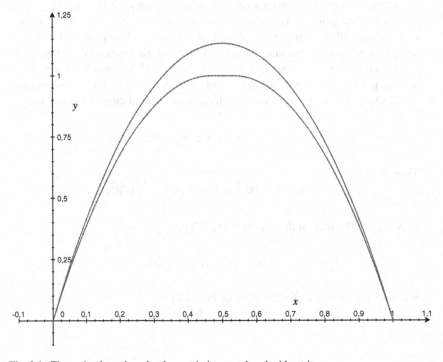

Fig. 9.1 The optimal graph under the restriction $u \leq 1$ and without it

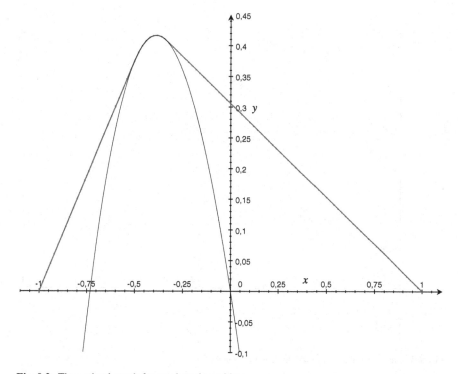

Fig. 9.2 The optimal graph for an obstacle problem

$$x(t) = \begin{cases} 2ut, & 0 < t < 1/2, \\ 1 + 2(1-u)(t-1), & 1/2 < t < 1, \end{cases}$$

its cost turns out to be

$$2u^2 + \frac{1}{6}u^2 + 2(1-u)^2 + \frac{1}{6}(1 + u + u^2).$$

The minimum value of this function of u corresponds to $u = 23/68 = 0.3382$, while the value at $t = 1/2$ of the true minimizer (which is easily found) is 0.4434.

9.5.2 Computer Exercises

1. As ε becomes smaller and smaller, somehow the optimal solution must look more and more like the function $u(x) = x/2$. However end-point conditions force to always have $u(0) = 0.5$, $u(1) = 0$, and so optimal solutions show a rapid

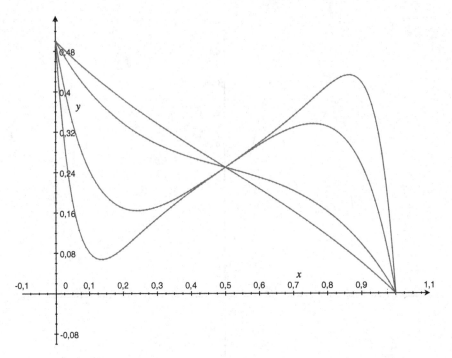

Fig. 9.3 The optimal solution for decreasing values of ε

transition at both end-points as Fig. 9.3 depicts. The smallest value of ε plotted is $\varepsilon = 0.002$.

2. The approximated solution is shown in Fig. 9.4. It looks a similar behavior as in the previous exercise: the optimal profile is like $u(x) = 0.5$ however feasible boundary data are incompatible, and that is way we see those two abrupt boundary layers. The solution of the associated Euler–Lagrange equation is precisely $u(x) = 0.5$.

3. Check Fig. 9.5 for optimal profiles for some increasing values of n.

4. It is elementary to reduce the problem to

$$\text{Minimize in } z(t): \quad \int_0^1 z''(x)^2 \, dx$$

under $z(0) = 0, z(1) = 1, z'(0) = z'(1) = 0$. The optimal solution is easily found to be $(z'(x) =)y(x) = 6x(1 - x)$. The numerical approximation, retaining the unknown pair (z, y) to keep the problem first-order, does not converge smoothly. Check Fig. 9.6 for a comparison. The reason for these difficulties are related to the fact that the auxiliary variational problem for each step in our iterative process is of the form

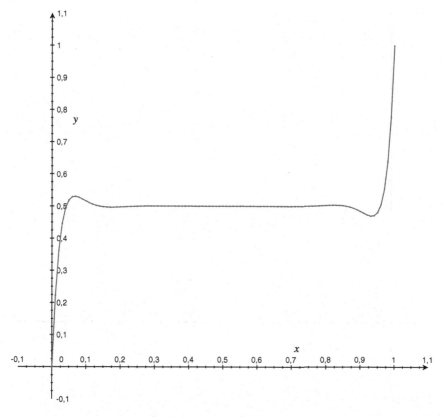

Fig. 9.4 Two abrupt boundary layers

$$\int_0^1 \left[y'(x)^2 + e^{v(x)(z'(x)-y(x))} + e^{w(x)(y(x)-z'(x))} \right] dx,$$

for $v(x)$, and $w(x)$, positive, and the joint, strict convexity of the integrand with respect to the triplet (y, y', z') is not correct. The situation is asking for further investigation. If we replace the integrand $y'(x)^2$ by $y'(x)^4$, and we, again, introduce the new variable $z'(x) = y(x)$, then the Euler–Lagrange equation for the resulting second-order problem is

$$[z''(x)^3]'' \text{ in } (0, 1), \quad z(0) = z'(0) = z'(1) = 0, z(1) = 1.$$

With a bit of care in the computations, one finds

$$z'(x) = \frac{3}{4c}(cx + d)^{4/3} - \frac{3}{4c}d^{4/3}.$$

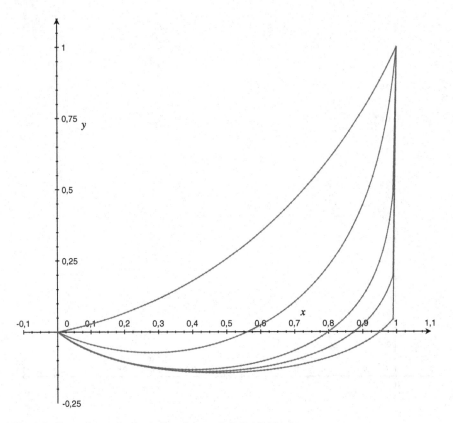

Fig. 9.5 Successive optimal solutions for $n = 2, 5, 8, 13, 15$

This function already complies with $z'(0) = 0$. If we insist in having $z'(1) = 0$, then there is no choice but $z''(x)$ to be constant. This is however incompatible with all of the end-point conditions. There is something to be understood here, but it is deferred to a more advanced course.

5. Check Fig. 9.7 for the minimizer for $p = 2$, $p = 4$, and $p = 8$. The bigger p is, the flatter the minimizer.

6. The line segment joining the initial and final points turns out to be critical (non-minimizer) por the functional perturbed with the corresponding exponentials and a constant multiplier. Therefore, the algorithm cannot escape from it. For a different initial path (or a non-constant multiplier), then we get the solution shown in Figs. 9.8 and 9.9.

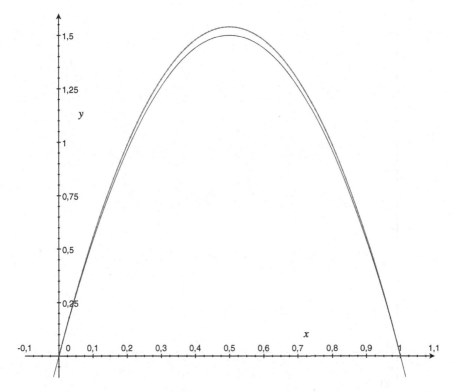

Fig. 9.6 Mismatch between approximation and exact solution

9.5.3 Case Studies

1. See Fig. 9.10 for two optimal trajectories.
2. Figure 9.11 plots two trajectories for the standard harmonic oscillator for two different radii. Integration time in both cases is $T = 50$ so that each integral curve is run several times around. Notice that there is essentially no mismatch on the various cycles around.
3. This is a typical example of an obstacle problem just as in the examples examined in the chapter. See Fig. 9.12 for the optimal profile for the symmetric situation, and Fig. 9.13 for the non-symmetric situation.
4. This is Newton solid of revolution. Figure 9.14 plots the optimal profile for several values of the constant M. The solid itself can be figured out by rotating the corresponding profile around the vertical axis.

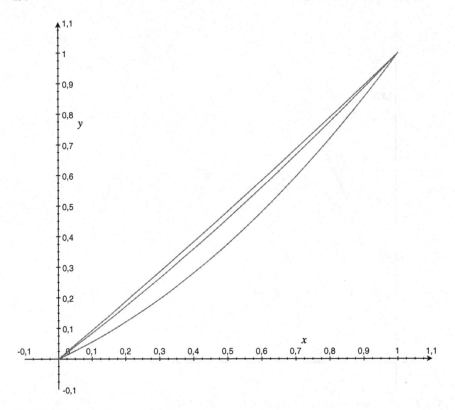

Fig. 9.7 Minimizers for $p = 2, 4, 8$

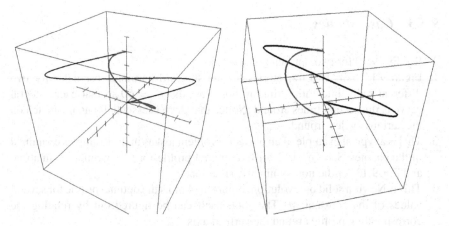

Fig. 9.8 Two views of the optimal solution (*blue*) and the initial curve (*black*) for the steering problem (color figure online)

Fig. 9.9 A final view from
above for the optimal path
for the steering problem

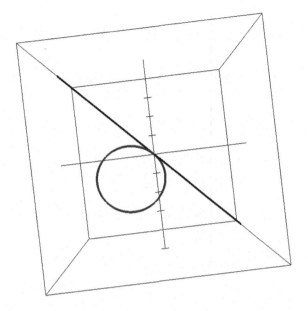

9.6 Chapter 7

9.6.1 Exercises to Support the Main Concepts

1. After eliminating the control variable from the formulation, we face the problem

$$\text{Minimize in } x(t): \quad \frac{1}{2} \int_0^1 (x'(t) - x(t))^2 \, dt$$

under the constraints $x(0) = 1$, $x(1) = 0$. The optimal state is

$$x(t) = \frac{e^{1-t} - e^{t-1}}{e - e^{-1}},$$

and the optimal control

$$u(t) = x'(t) - x(t) = \frac{2e^{1-t}}{e^{-1} - e}.$$

2. This is just Pontryaguin's maximum principle.
3. If we do not restrict the size of U in any way, then there is always a constant
   such vector to take any initial condition x_0 into any other terminal condition x_T.
   Indeed, the exponential formula yields

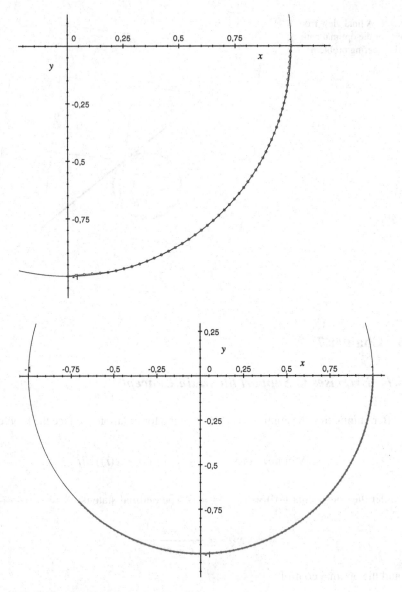

Fig. 9.10 Two optimal, constrained trajectories

$$\mathbf{x}(T) = e^{T\mathbf{A}}\mathbf{x}_0 + e^{T\mathbf{A}} \int_0^T e^{-s\mathbf{A}}\,ds\,\mathbf{U},$$

and so

$$\mathbf{U} = \left(e^{T\mathbf{A}} \int_0^T e^{-s\mathbf{A}}\right)^{-1} (\mathbf{x}_T - e^{T\mathbf{A}}\mathbf{x}_0).$$

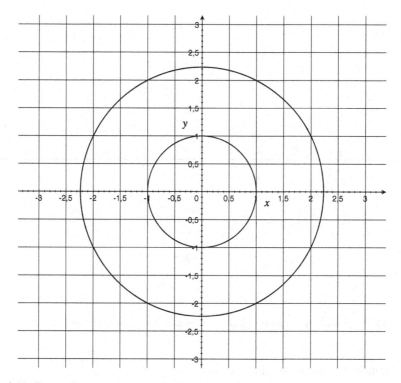

Fig. 9.11 Two optimal trajectories for the harmonic oscillator

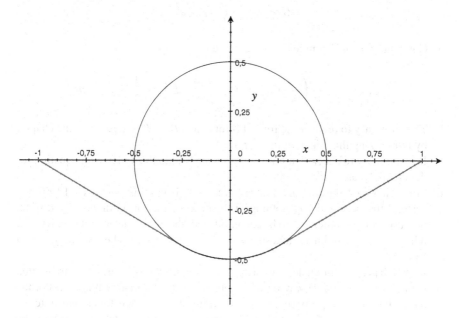

Fig. 9.12 An elastic string supporting a circular obstacle: symmetric case

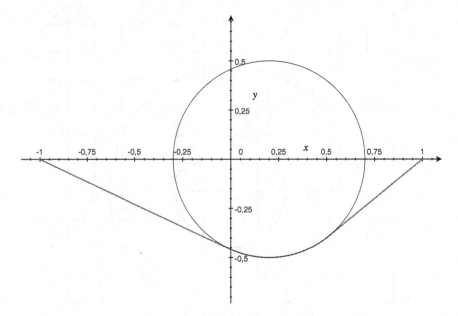

Fig. 9.13 An elastic string supporting a circular obstacle: non-symmetric case

For the one-dimensional situation, we get

$$U = \frac{a}{e^{Ta} - 1}(x_T - e^{Ta}x_0).$$

4. Computations in Example 7.6 directly yield

$$E(t_0, t_1) = \frac{1}{2}\left(-\frac{a}{2}t_1^2 + bt_0t_1 + \frac{b}{2}t_0^2 - L\right)^2 + \frac{1}{2}(-at_1 + bt_0)^2.$$

Then it is easy to recover the formulas for t_0 and $t_1 = T - t_0$ given in the chapter by minimizing this function in (t_0, t_1).

5. It is very easy to check that both perspectives lead to the underlying equation for the optimal state $-x'' + x = 0$.

6. In the first case the optimal strategy is $u \equiv 0$. If nothing is expected from the system, the best is doing nothing. The second situation is drastically distinct because there is something to be accomplished: the final state cannot be anything, it has to be some point x_T, and so, as soon as x_T is not x_1, the best strategy cannot be $u \equiv 0$.

7. Dynamic programming deals with optimization problems where costs are written naturally in a recursive way because some of its terms arise in this way. For the particular situation proposed, it is easy to find that the cost functional in terms of (u_0, u_1) is $2u_0 + 6u_0u_1 + u_1$ with a minimum value -7 taken on at $(-1, 1)$.

Fig. 9.14 Optimal profiles
for Newton's solid of
revolution for several values
of $M = 1, 2, 3, 4$

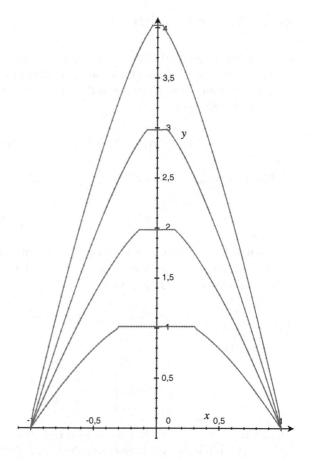

Equivalently, the cost functional can be reinterpreted in terms of (x_1, x_2). It turns
out to be

$$\frac{2}{3}x_1 + 2x_2 - \frac{x_2}{x_1 + 1}$$

over the feasible set

$$|x_1| \leq 3, \quad |x_2| \leq |x_1 + 1|.$$

It is instructive to plot this set. The point of minimum is achieved at $(-3, -2)$
which corresponds exactly with the previous pair.

9.6.2 *Practice Exercises*

1. Through the hamiltonian, it is easy to conclude that both u and the second multiplier must be solutions of the same equation of the harmonic oscillator. In this way the optimal state has to be of the form, taking into account the initial and terminal conditions,

$$x(t) = 2\cos t + \frac{2}{\pi}(\sin t - t\cos t).$$

The optimal strategy is then

$$u(t) = x''(t) + x(t) = \frac{4}{\pi}\sin t.$$

2. The hamiltonian is linear in u, and so the optimal strategy is of type bang-bang. The switching function, after using the differential equation for the costate $p(t)$ and the transversality condition, is

$$1 - p(t) = 2 - e^{1-t}.$$

Hence, the optimal strategy is

$$u(t) = \begin{cases} 1, & 0 < t < 1 - \log 2 \\ -1, & 1 - \log 2 < t < 1. \end{cases}$$

3. (a) If you bear in mind the transversality condition $p_1(2) = 0$, precisely because $x(2)$ is not prescribed, and using Pontryaguin's principle, one finds the optimal triplet

$$x(t) = -\frac{t^4}{24} + \frac{t^3}{3} - \frac{11}{12}t^2 + t, \quad u(t) = -p_2(t) = -\frac{t^2}{2} + 2t - \frac{11}{6}.$$

The distance run is $x(2) = 1/3$.

   (b) The optimal strategy is the same because the optimal $u(t)$ from the previous item, turns out to remain in the interval $[-2, 2]$ for all $t \in [0, 2]$.

   (c) We would need to find the two extreme (maximum and minimum) of the optimal strategy $u(t)$ in the interval $[0, 2]$. The optimal interval is $[-11/6, 1/6]$. As soon as the feasibility interval for u does not entirely contain this concrete interval, the optimal strategy would have to change.

4. We are asking about having $x(t) \equiv 1$ for all times $t \in [0, 1]$. According to the dynamic law, this is possible if $0 = 1 + u$, i.e. $u = -1$, but this is not feasible. Nevertheless, we can minimize the (square of the) distance from the final state $x(1)$ to the desired state 1. The hamiltonian would be $H = p(x + u)$. It is easy to conclude that the optimal strategy is constant at one of the two end-point values

$u = -1/2$ or $u = 1$. Comparing the cost of both possibilities, one concludes that $u = -1/2$ is the optimal strategy. As indicated above, if $u \equiv -1$ is eligible, then the optimal strategy would be this one.

5. Note how the term 1 does not interact with the optimization process so that we can ignore it altogether. If we keep in mind the transversality condition $p_1(2) = 0$, Pontryaguin's principle takes us to the fact that the optimal strategy u is independent of time. End-point conditions for x imply that $u = -1/2$ is optimal, and then $x(t) = t - t^2/4$ is the optimal evolution of distance with $x(2) = 1$.

6. (a) If $u(t) = U$, constant, then $x(t) = -U + (100 + U)e^t$. Taking this expression into the cost functional, we realize that it is quadratic in U, and so it reduces to the expression $AU^2 + BU + C$ with

$$A = \frac{1}{2}(e^{10} - 1)^2 + \int_0^{10} [1 + 3(e^t - 1)^2] \, dt,$$

$$B = 100e^{10}(e^{10} - 1) + 600 \int_0^{10} e^t(e^t - 1) \, dt,$$

$$C = 5000e^{20} + 30000 \int_0^{10} e^{2t} \, dt.$$

   (b) There is no special difficulty in determining the optimal $u(t)$ under no restriction on its size, and the transversality condition $p(10) = x(10)$. The optimal strategy turns out to be

$$u(t) = -\frac{20e^{-t}}{1 + 3e^{-20}}.$$

7. For $u = 0$, it is easy to find that the cost is $e - 1$. For the cost in the second part, more calculations are involved, but after a bit of patience solving one problem after the other, one finds that the cost is $2e^{1/2} - 1$. Given that $2 > \sqrt{e}$, this second price is larger than the first one. The optimal strategy requires to solve the problem

$$\min_{u \in [-1,1]} (u^2 - pu)$$

for the costate p, which turns out to be $p(t) = e^{1-t} - 1$. A careful analysis through Pontryaguin's principle leads us to the optimal strategy

$$u(t) = \frac{1}{2}(e^{1-t} - 1)$$

because this quantity always belongs to $[0, 1]$ for $t \in [0, 1]$. The corresponding state is

$$x(t) = -\frac{1}{2} + \frac{e^{1-t}}{4} + \left(\frac{3}{2} - \frac{e}{4}\right)e^t.$$

8. The control $u \equiv 0$ will be feasible if it is capable of taking the system back from $x(0) = 2$ to $x(1) = 2$. It is not so. Suppose we seek a constant control u. Then we need to determine such constant, together with the constant of integration so that

$$x' = -x + u, \quad x(0) = x(1) = 2.$$

If we take $u = 2$, evidently the constant solution $x \equiv 2$ does that. Since the cost functional is

$$\int_0^1 x(t)\, dt,$$

through the hamiltonian, the costate is of the form $p(t) = ce^t + 1$ for some constant. This function can, at most, change signs in $(0, 1)$ once, and so one tries a strategy of the form $u(t) = 2$ for $t \in (0, t_0)$, $u(t) = -2$ for $t \in (t_0, 1)$ (why is not the other possibility admissible?). Examining the two successive problems, one concludes that in fact $t_0 = 1$, and the optimal strategy is precisely $u(t) \equiv 2$.

9.6.3 Case Studies

1. It is a minimal time problem. The strategy is of bang-bang type taken on the two extreme values $-U$ and U. The multiplier is of the form $p(t) = ce^{kt}$, and so its sign is determined by the sign of the constant c, but there cannot be a change of sign, and hence the optimal strategy is constant at either $-U$ or U. Given that a reduction of the population is sought, it is clear that $u \equiv -U$. The population $x(t)$ is found to be

$$x(t) = \left(P + \frac{U}{k}\right) e^{-kt} - \frac{U}{k}.$$

Imposing the condition $x(T) = P/2$, we arrive at

$$T = \frac{1}{k} \log\left(1 + \frac{Pk}{Pk + 2U}\right).$$

If a constant population $x \equiv P/2$ is wished to be maintained, then the state law needs to permit $0 = -kP/2 + u$, and so the control $u = kP/2$ must be eligible. This requires $kP/2 \le U$. Otherwise, it is not possible.

2. The hamiltonian of the problem is linear in u, and so the optimal strategy is of bang-bang type. The costate is $p(t) = ce^{\alpha t}$ for a unknown constant c. The sign of this constant carries us to the two extreme possibilities $u \equiv -U$ or $u \equiv 0$. Evidently, $u \equiv 0$ would do nothing, and so $u \equiv -U$ is optimal. In such a case the level of gas will evolve according to

$$x(t) = \frac{U}{\alpha} + \left(C - \frac{U}{\alpha}\right) e^{\alpha t}.$$

A null level will be achieved, provided $C < U/\alpha$, when

$$T = \frac{1}{\alpha} \log \frac{U}{U - \alpha C}.$$

If $C \geq U/\alpha$, the null level is impossible to achieve in a finite time.

3. Optimality conditions take us to the equation

$$x''(t) + w_0^2 x(t) = cos(w_0 t) + a \cos(w_0 t) + b \sin(w_0 t)$$

for some constants a and b. The formal, general solution is of the form

$$x(t) = A \cos(w_0 t) + B \sin(w_0 t) + \alpha t \cos(w_0 t) + \beta t \sin(w_0 t)$$

and constants A, B, α and β are to be determined by demanding

$$x(0) = x_0, x'(0) = x_0', \quad x(\pi/w_0) = x'(\pi/w_0) = 0.$$

Careful computations yield

$$A = x_0, \quad B = \frac{x_0}{\pi} + \frac{x_0'}{w_0}, \quad \alpha = -\frac{w_0 x_0}{\pi}, \quad \beta = -\frac{x_0'}{\pi}.$$

The optimal $u(t)$ would be

$$u(t) = x''(t) + w_0^2 x(t) - \cos(w_0 t)$$
$$= \frac{2 w_0^2 x_0}{\pi} \sin(w_0 t) - \left(\frac{2 x_0' w_0}{\pi} + 1\right) \cos(w_0 t).$$

4. There is no special difficulty in finding the optimal solution through optimality, except to avoid confusion with notation. There is no formal difference with respect to the number of components of the control variable. The optimal solution is given through the identity

$$\mathbf{u}(t) = -e^{(t-T)\mathbf{A}^*} \mathbf{A}^* (\mathbf{A}\mathbf{x}(T) + \mathbf{b})\mathbf{B},$$

where $\mathbf{A}^*$ is the transpose, and $e^{\mathbf{M}}$ is the exponential of the matrix $\mathbf{M}$.

5. The rule for the optimal solution u is, through Pontryaguin's principle,

$$u = \begin{cases} 0, & \text{if } p \leq 0, \\ p, & \text{if } 0 \leq p \leq 2, \\ 2, & \text{if } p \geq 0. \end{cases}$$

The dynamical law for the multiplier p yields $p(t) = ce^t$ for some constant c, which, according to the interpretation of the problem, must be positive. If we put $t_0 = \log(2/c)$, then we will have

$$u(t) = \begin{cases} ce^t, & 0 \le t \le t_0, \\ 2, & t_0 \le t \le T. \end{cases}$$

We finally proceed to solve the state law in two successive steps according to the two regimes for $u(t)$. Careful explicit computations lead to

$$u(t) = (e^T + \sqrt{e^{2T} - 4})e^t, \quad t_0 = \log\left(\frac{2}{e^T + \sqrt{e^{2T} - 4}}\right).$$

Notice that this solution requires $T > \log 2$, for otherwise the objective $x(T) = 0$ is impossible to achieve, unless the maximum dosage rate is increased.

6. If $x(t)$ and $y(t)$ represent positions of the police car and the motorist, we will have

$$y(t) = tv, \quad x''(t) = u(t), \quad u \in [-b, a],$$

and T ought to be determined so that

$$x(T) = y(T) = Tv, \quad x'(T) = y'(T) = v,$$

with T as small as possible. The hamiltonian is linear in u, and so the optimal strategy is of bang-bang type. The coefficient in front of u in the hamiltonian is $p_2(t)$ which happens to be affine in t: $p_2(t) = ct + C$. There is, accordingly, at most one change in the use of u: from a to $-b$. Computations resolving the state law in two successive steps, take us to the system

$$v = -b(T - t_0) + at_0, \quad Tv = -\frac{b}{2}(T - t_0)^2 + at_0(T - t_0) + \frac{a}{2}t_0^2.$$

Some careful calculations lead to

$$t_0 = \frac{2v}{a}, \quad T = \frac{2v}{a} + \frac{v}{b}.$$

7. If we do nothing $u \equiv 0$, it is easy to arrive at the fact that the crop will vanish after one unit of time. For $u = 1$, the extinction time $t_0 = \log 3 > 1$. Through optimality and the interpretation of the problem, there is no difficulty in having that the optimal strategy is $u(t) = 2$ for $0 \le t \le \bar{t}$, and $u(t) = 0$ for $\bar{t} \le t \le T$ for some switching time $\bar{t}$. Indeed, the earliest that the predator can be eliminated is $\bar{t} = \log 2$, by setting $u \equiv 2$, with the crop left at $x(\log 2) = \log 4 - 1 > 0$. Once $y \equiv 0$, set $u \equiv 0$, and let the crop reach $1/2$ in the appropriate time.

8. This is a problem with an unconstrained control. There is no special difficulty, but it requires some patience, in setting optimality conditions and solving successively the differential equation for multipliers and states. The adjustment of arbitrary constants also demands some careful arithmetic. The optimal pair is

$$u(t) = -Ce^t - c, \quad x_1(t) = a - Ae^{-t} - ct - \frac{C}{2}e^t,$$

with

$$A = \frac{e(0.1e - 0.15)}{4e - 3 - e^2}, \quad a = \frac{2}{e}A,$$
$$C = (4e^{-1} - 2)A - 0.2, \quad c = 0.1e + (e^{-1} - 2 + e)A.$$

Approximate numerical values are

$$A = 0.68416, \quad a = 0.5034, \quad C = -0.5616, \quad c = 1,015.$$

It is instructive to plot both $u(t)$ and $x_1(t)$ for $t \in [0, 1]$.

9. Though one can use Pontryaguin's principle to find the optimal strategy, the interpretation of the problem directly leads to the best choice $u(t) \equiv 1$. The largest voltage difference is

$$i(T) = \frac{1}{R}(1 - e^{-RT/L}).$$

9.7 Chapter 8

9.7.1 Exercises to Support the Main Concepts

1. The integrand for the first problem is

$$F(x, \xi) = \frac{1}{2}(\xi - x)^2,$$

so that the equivalent variational problem is

$$\text{Minimize in } x(t): \quad \int_0^1 F(x(t), x'(t)) \, dt$$

subject to $x(0) = 1, x(1) = 0$. For the second problem we would have

$$F(x, \xi) = \begin{cases} \frac{1}{2}(\xi - x)^2, & |\xi - x| \leq 1, \\ +\infty, & |\xi - x| > 1. \end{cases}$$

2. The integrand is

$$F(x, \xi) = \begin{cases} \frac{1}{8} \min\{(-x + \sqrt{x^2 + 4\xi})^2, (-x - \sqrt{x^2 + 4\xi})^2\}, & x^2 + 4\xi \geq 0, \\ +\infty, & x^2 + 4\xi < 0. \end{cases}$$

9.7.2   Computer Exercises

1 (a) The optimal pair is

$$u(t) = \frac{x_0}{1 + e^{4T}} e^{2t} - \frac{3e^{4T} x_0}{1 + e^{4T}} e^{-2t},$$

$$x(t) = \frac{x_0}{1 + e^{4T}} e^{2t} + \frac{e^{4T} x_0}{1 + e^{4T}} e^{-2t}.$$

(b) For the numerical approximation, we eliminate the control variable from the formulation of the problem, and rewrite the term

$$\frac{1}{2} x^2(2) = \frac{1}{2} + \int_0^2 \frac{1}{2} \frac{d}{dt} [x^2(t)] \, dt = \int_0^2 x(t) x'(t) \, dt$$

$$= \int_0^2 x(t)(x(t) + u(t)) \, dt.$$

We can ignore the constant $1/2$ as it does not interfere with optimization, and so, after eliminating the control variable $u = x' - x$ from the formulation, we face the problem

$$\text{Minimize in } x(t): \quad \int_0^2 \frac{1}{2} [4x(t)^2 + x'(t)^2] \, dt$$

under the sole condition $x(0) = 1$. Once the optimal $x(t)$ is found or approximated, the optimal u is given by $x'(t) - x(t)$. Compare the exact and approximated solutions for estate and control in Fig. 9.15.

2. The non-friction case has been examined in the corresponding chapter. A comparison for the friction and non-friction cases for horizon $T = 5.03$ which is, approximately, the minimal time for the non-friction case can be seen in Fig. 9.16, while the minimum-time curve for the friction case is plotted in Fig. 9.17.

3. This is a standard quadratic control problem with no constraint for control. It can be transformed easily into a constrained variational problem in the form

$$\text{Minimize in } (x_1, x_2): \quad \frac{1}{2} \int_0^{10} \left[\beta x_1(t)^2 + r(x_2'(t) + x_2(t))^2 \right] dt$$

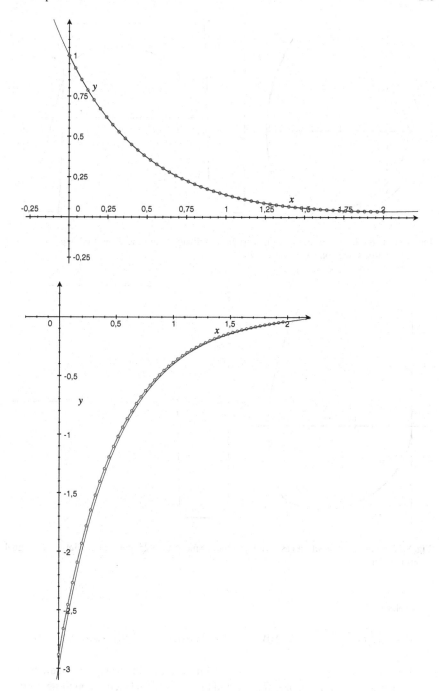

Fig. 9.15 Comparison between exact and approximated solutions: upper, state; lower, control

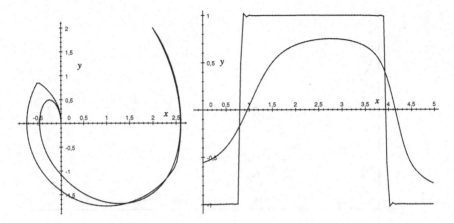

Fig. 9.16 Paths for the friction and non-friction oscillator with time $T = 5.03$ and initial point $(2, 2)$. Corresponding control, lower picture

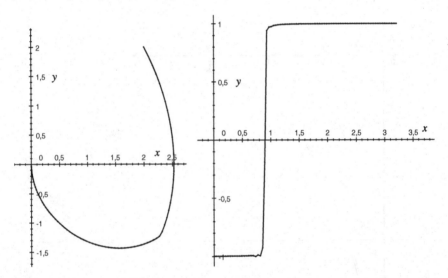

Fig. 9.17 Path for the friction case with minimum time $T = 3.25$ and initial point $(2, 2)$. Signal control, down

under

$$x_1'(t) = x_2(t) \text{ in } (0, 10), \quad x_1(0) = x_2(0) = 1, x_1(10) = x_2(10) = 0.$$

The curves in Fig. 9.18 show the optimal trajectories in the $x_1 - x_2$ plane for the three cases: red for (a), blue for (b), and black for (c). The corresponding optimal controls are plotted on the right.

4. This example is a bit peculiar in that the final state is not fully determined but velocity is free, and this has to be incorporated into the approximation algorithm. Other than that is standard, namely,

$$\text{Minimize in } (x_1, x_2) : \quad \frac{1}{2} \int_0^1 [x_1'(t)^2 + x_2'(t)^2] \, dt$$

such that

$$x_1'(t) = x_2(t) \text{ in } (0, 1), \quad x_1(0) = x_2(0) = 0, x_1(1) = 1.$$

Both the optimal control signal (in black), and space x_1 (in blue) can be seen in Fig. 9.19.

5. The approximated solutions for Duffin's equation are shown in Figs. 9.20 and 9.21. Curves in black correspond to Duffin's while blue ones are for the linear case without the cubic perturbation.

6. This exercise is setup exactly as it has been considered in the chapter but playing with the time T in which the oscillator brings the state from $(2, 2)$ to $(0, 0)$. In Fig. 9.22, one can see a plot of the optimal paths and signal controls for times $T = 8, 6, 5.5, 5.25, 5.125, 5.0625,$ and 5.03125.

9.7.3 Case Studies

1. The equivalent variational problem is easily set up in the following terms

$$\text{Minimize in } x(t) : \quad \int_0^T (1 - x(t) - x'(t))^2 \, dt$$

subject to

$$-1 - x(t) - x'(t) \le 0, -1 + x(t) + x'(t) \le 0 \text{ in } (0, T), \quad x(0) = 1, x(T) = 0.$$

Optimal dosage rate (in black), and the corresponding evolution of acid uric in blood (in blue) can be checked out at Fig. 9.23 for decreasing values of $T = 10, 5, 2, 1, 0.8$.

2. This is a typical third-order example. We use variables

$$x_1 = \theta, \quad x_2 = \theta', \quad x_3 = \alpha,$$

so that we need to enforce

$$x_1'(t) = x_2(t), \quad x_2'(t) = -x_2(t) - x_3(t), \quad x_3'(t) = -x_3(t) + u(t),$$

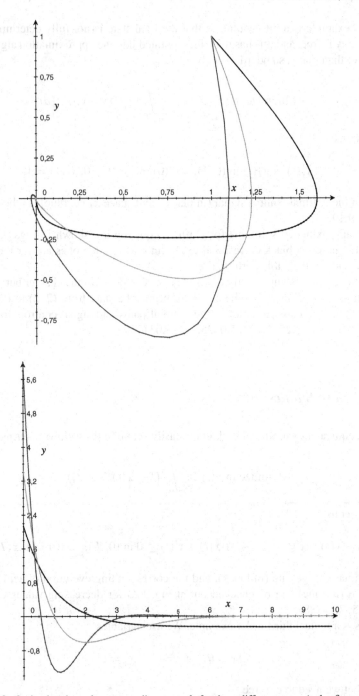

Fig. 9.18 Optimal paths and corresponding controls for three different cases (color figure online)

Fig. 9.19 Optimal path and corresponding control signal for a double integrator with a quadratic cost (color figure online)

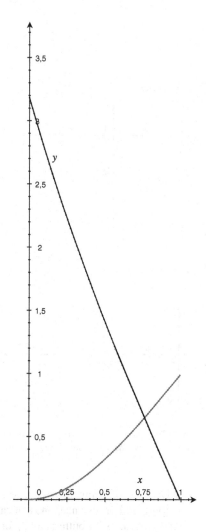

together with appropriate initial and final condition for the system. In terms of inequalities we would put

$$x_1'(t) - x_2(t) \le 0, \quad x_2(t) - x_1'(t) \le 0,$$
$$x_2'(t) + x_2(t) + x_3(t) \le 0, \quad -x_2'(t) - x_2(t) - x_3(t) \le 0,$$
$$x_3'(t) + x_3(t) - 1 \le 0, \quad -1 - x_3'(t) - x_3(t) \le 0.$$

Check on Fig. 9.24 those two optimal trajectories with their corresponding optimal control signals.

3. This is a typical example to see the basic principle of dynamic programming Principle 8.2 in action, even though the size of the problem is small and the

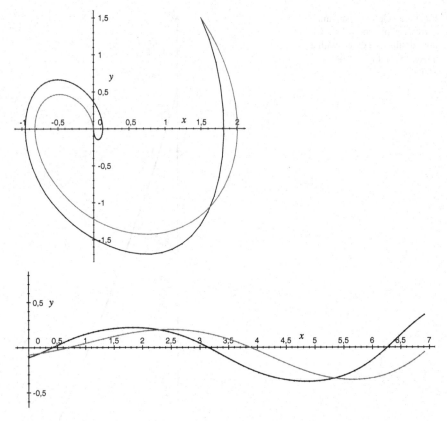

Fig. 9.20 Comparison between the two situations with and without the cubic term for Duffing's equation with time $T = 7$ and initial point (1.5, 1.5) (color figure online)

advantage of using it against a direct, brute-force approach may not look so important. For instance, one can easily compute the benefit of utilizing 1 unit for cheese; 2 units for butter; and 2, for yogurt. According to the data in the table, it is 34. This is better than taking the 5 units of milk for cheese, which would amount to a benefit of 32. The brute-force solution would cover all the possibilities and check on the one providing a maximum benefit. Dynamic programming also aims at that, but it does organize possibilities in a more rational way.

We will organize the successive steps for this situation according to the number of products that are produced: one only product to which the five units of milk are destined; two different products with two distinct possibilities: either four units to one, and one to the other, or three units to the first and two to the second; three products with alternatives: three, one, one or two, two, one; and finally, four different products with the unique possible distribution: two, one, one, one. Based on the data of the provided table, it is easily seen that the maximum values for those various steps are, respectively, 32, 37, 38, 39, and 37, corresponding

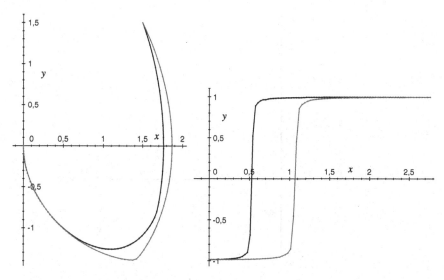

Fig. 9.21 Paths for optimal time for Duffing's equation with time $T = 2.55$, and without the cubic term $T = 2.8$ (color figure online)

the maximum value 39 to the optimal strategy with two units for cheese, one for cottage cheese, and two for yogurt.

4. The solution is 17 corresponding to the distribution 1 million for youth, 2 for culture, and another 2 for environment.

5. In terms of the four component x_i, $i = 1, 2, 3, 4$ of $\mathbf{x}$, the equations of the state system are

$$x_1' = x_2, \quad x_2' = -\frac{c}{Jn^2}x_1 - \frac{v}{J}x_2 + \frac{c}{Jn^2}x_3 + \frac{k}{Jn}u,$$

$$x_3' = x_4, \quad x_4' = \frac{c}{ml^2}x_1 - \frac{c}{ml^2}x_3.$$

For the specific values of the parameters, the corresponding constrained variational problem would be associated with the integrand

$$e^{y_1(0.035x_1' + 0.036x_1 + 0.25x_2 - 0.036x_3 - 1.7)}$$

$$+ e^{y_2(-0.035x_2' - 0.036x_1 - 0.25x_2 + 0.036x_3 - 1.8)}$$

$$+ e^{y_3(x_1' - x_2)} + e^{y_4(x_2 - x_1')} + e^{y_5(x_3' - x_4)} + e^{y_5(x_4 - x_3')}$$

$$+ e^{y_7(0.111x_4' - x_1 + x_3)} + e^{y_8(-0.111x_4' + x_1 - x_3)}.$$

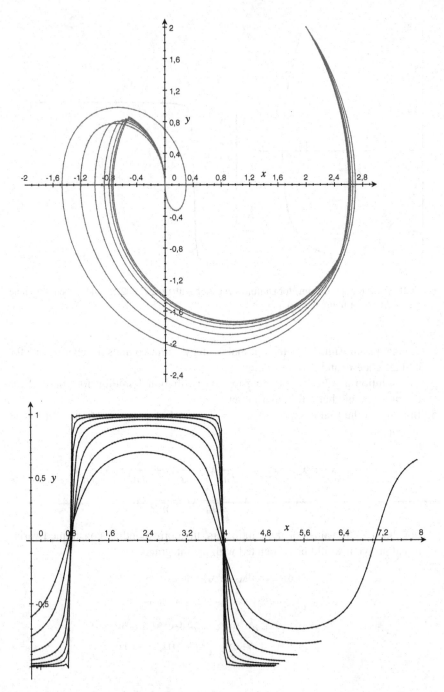

Fig. 9.22 Optimal paths and corresponding optimal signal controls (*down*) for the classical harmonic oscillator and $T = 8, 6, 5.5, 5.25, 5.125, 5.0625, 5.03125$

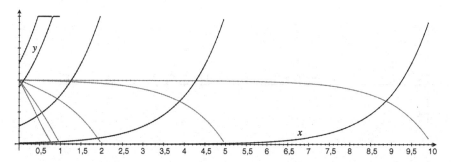

Fig. 9.23 Optimal dosage strategies (*black*) and evolution of acid uric levels (*blue*) for various values of $T = 10, 5, 2, 1, 0.8$ (color figure online)

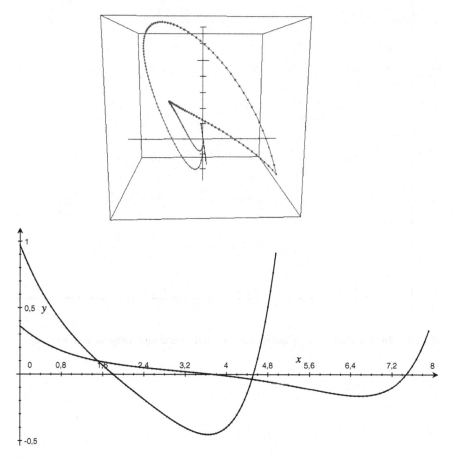

Fig. 9.24 Optimal paths (*up*) and optimal signals (*down*) for a third-order problem with $T = 8, 5$

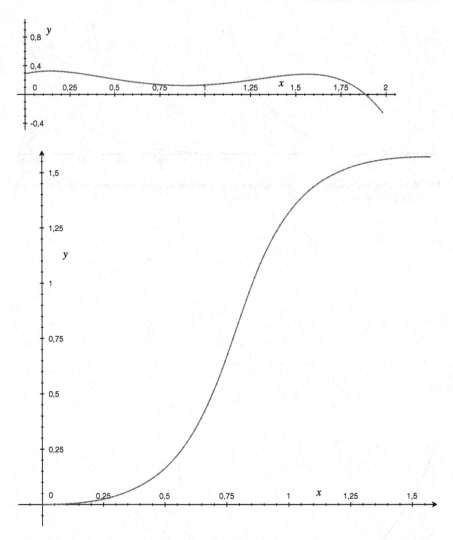

Fig. 9.25 The control signal (*up*), and the state (*down*, first and third components) for $T = 2$

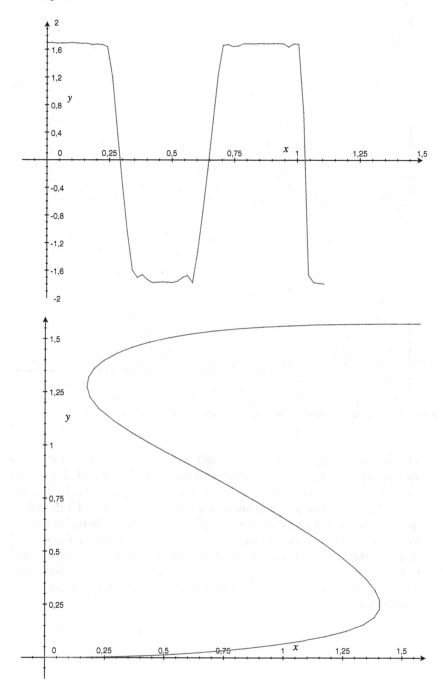

Fig. 9.26 The control signal (*up*), and the state (*down*, first and third components) for $T = 1.125$

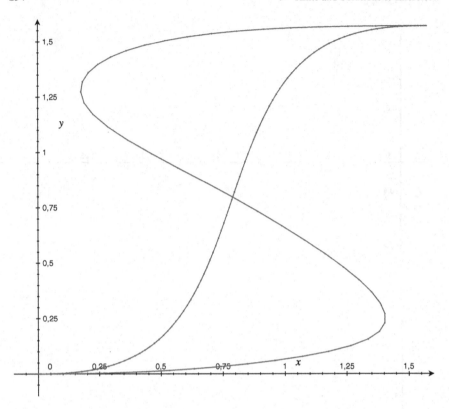

Fig. 9.27 A comparison between the evolution of the state for $T = 2$ and $T = 1.125$

The variables y_j are the auxiliary multipliers. Figures 9.25 and 9.26 show the approximated signal control and evolution of the state (first and third components, the two angles describing the system) for the two extreme cases $T = 2$ and $T = 1.125$, respectively. The minimal time is very close to $T = 1.125$, and we see how the optimal signal control is getting pretty close to being an optimal bang-bang solution. If one tries with $T = 1.1$, because the minimal time is larger than this value of T, the algorithm cannot and does not converge. Finally, Fig. 9.27 plots, for comparison purposes, the evolution of the state for those two extreme, proposed cases. We clearly see how much more aggressive is the evolution for a smaller time horizon.

Printed in the United States
By Bookmasters